TROPICAL REFRIGERATION AND AIR-CONDITIONING

Longman Industrial Crafts Series
(General Editor: Roger L. Timings)

TROPICAL REFRIGERATION AND AIR-CONDITIONING

L. W. Cottell and S. Olarewaju

Longman
London and New York

Longman Group Limited,
Longman House,
Burnt Mill, Harlow,
Essex CM20 2JE,
England
and Associated Companies
throughout the World.

First published in 1983
© Longman Group Limited 1983

All rights reserved. No part of this
publication may be reproduced, stored
in a retrieval system or transmitted
in any form or by any means, electronic,
mechanical, photocopying, recording or
otherwise without the prior permission
of the Copyright owner.

ISBN 0 582 65803 9

Set in 10/11pt Helvetica

Illustrations by Multiplex Techniques Ltd.

Printed in Singapore by
Huntsmen Offset Printing (Pte) Ltd

British Library Cataloguing in Publication Data

Cottell, L. W.
  Tropical refrigeration and
  air-conditioning.
  — (Longman industrial crafts series)
  1. Air conditioning – Equipment and
  supplies
  2. Refrigeration and refrigerating
  machinery
  — Tropics
  I. Title II. Olarewaju, S.
  697.9.3   TH7687
  ISBN 0-582-65803-9

Library of Congress Cataloging in Publication Data

Cottell, L. W.
  Tropical refrigeration and air-conditioning
  (Longman industrial crafts series)
  Includes index.
  1. Refrigeration and refrigerating
  machinery 2. Air conditioning.
  I. Olarewaju, S. II. Title III. Series.
  TP492.C73 1983 621.5 82–16215
  ISBN 0-582-65803-9

Contents

Preface .... *page* vii

1
The history of refrigeration 1
The pioneers, Breakthrough, The big boom.

2
Heat, temperature and pressure 4
Heat, Temperature, Specific heat capacity, Units of heat, Sensible and latent heat, Heat transfer, Pressure, Pressure/temperature relationships, Theoretical refrigeration cycle, The properties of air.

3
Electricity – theory 17
The electron theory, How currents are generated, Basic principles of current electricity, Magnetism and electricity, Generators, Transformers, Capacitance, Electrical motors, Power distribution, Earthing, Colour coding of conductors, The Peltier effect.

4
Refrigerants 35
Storage and safety, Refrigerant properties, Pressure/temperature relationships, Refrigerant leak detection, Refrigeration oils, Secondary refrigerants.

5
Domestic appliances 46
Refrigerators and freezers, Domestic air-conditioners, Room units, Dehumidifiers, Installing domestic appliances, Maintaining domestic appliances.

6
Commercial and industrial refrigeration 71
Reach-in refrigerators, Cabinets and insulation, Condensing units, Compressors, Condensers, Liquid-line fittings, Refrigerant controls, Food storage, Refrigerated display cases, Cold rooms and cold storage, Refrigeration cooling loads, Block ice plant, Food freezing, Low temperature refrigeration systems, Refrigerant pumps.

7
Commercial and industrial air-conditioning equipment 110
Packaged comfort-cooling units, Close control air-conditioning, Central plant, Centrifugal chillers, Packaged water chillers, Water pumps, Airhandling units, Other air terminal units, Air distribution, Evaporative air coolers, Electronic air cleaners, Legionnaires' disease.

8
Pipeline and ductwork installation 138
Safety precautions, Notes on contents, Refrigerant pipelines, The importance of cleanliness, Evacuation and dehydration, Other leak test procedures, Refrigerant charging, Charging procedures, Freon piping practices, Water piping, Air distribution, Anti-vibration mountings and foundations, Drive kit installation.

9
Installation, commissioning and maintenance planning 164
Installation procedures, Planned maintenance, Things to avoid.

10
Trouble-shooting 168
Whole system (electrical), Self-contained refrigerators, Walk-in refrigerators, Airhandling units, Thermostatic expansion valves, Solenoid valves, Condensing units, Hot gas bypass valves, Water chillers, Water cooling towers, Water pumps, Electric motors.

v

11
Refrigeration system service *188*
Safety precautions, Adjustment of controls, Accessible and open compressors, Replacing welded hermetic compressors, Gas charging methods, Air or overcharge?, Pressures in welded hermetic systems, Workshop layout.

12
Electrical service procedures *199*
Safety precautions, Test equipment, Test sequences, Starting relays, Capacitor tests, Compressor motor tests, Three-phase motor tests, Other components, Conclusion.

13
First aid *210*
Electric shock, Heat burns, Cold burns, Heavy bleeding, First aid kits, First aid training.

14
Conversion factors *213*
Conversion table, Copper tubing dimensions, Steel tubing dimensions, Sheet metal gauges.

Glossary *217*

Practice papers *222*

Preface

In West Africa, a large population and a tropical climate combine to produce heavy demands for refrigeration and air-conditioning. The personnel required to install and to service this equipment are usually trained in accordance with the subject matter and examination policies defined by the West African Examination Council in their Syllabus 157. This is designed to be taught over three years, and to be followed by two examinations. The first is on the science of the trade, and the second concentrates on equipment and its installation and service.

A broadly similar syllabus is followed by students in some other African countries, the Caribbean, and the Near and Far East, who wish to sit one of the three examinations in 'R & A' conducted by the City and Guilds of London Institute. This is Syllabus 827.

Both examining bodies seek proof of practical rather than abstract knowledge, and require constant attention to safety precautions and sound engineering practices. Both instruction and examinations are based on types of equipment in widespread current use.

This book is designed as an aid to students of either syllabus. In accordance with syllabus requirements, theory is treated in a simple and straightforward fashion; but equipment and procedures are described in sufficient detail to enable students to work safely, efficiently, and in logical sequences. The text is designed to be of continuing value to practising engineers, in terms of fully current practices, and future rather than historic trends.

Since this branch of engineering frequently demands the ability to visualise exactly what is happening inside hermetically-sealed metal systems, we have been generous in our use of illustrations. The kind co-operation of a number of manufacturers, and in particular of ASHRAE, in authorising the reproduction of copyright diagrams and information has enabled us to secure a higher degree of detail than would otherwise have been possible. Acknowledgements are made elsewhere; but the authors would like to express their appreciation of their help.

As we write, metrication is progressing faster in some countries, and some branches of engineering, than in others. The units of measure quoted in various text books and catalogues have not yet reached a common standard, and some students may not yet realise that the older 'metric' and newer 'SI' systems are not identical. There is ample room for confusion; and we have therefore quoted SI in the text with British units in parentheses. However in cases where SI units are not used in practice they are omitted. Chapter 14 summarises preferred units of measure and gives conversion factors for all three systems. This includes notes on 'rounding off' practices in common use by specific trades, which can easily confuse the uninitiated.

To avoid the risk of students spending time which they can ill afford on subjects not included in the syllabus on which they personally will be examined, we have identified items required by only one of the two bodies:

(WA) .. West African Examination Council 157 only,

(CG) .. City & Guilds of London 827 only.

All are however relevant to the industry in which students have elected to earn their livings and there are inevitable overlaps of chapter or paragraph contents.

Sample examination papers are provided at the end of the book to familiarise students with their format and marking balance. Questions at the ends of chapters include some taken from past WAEC papers, and about 30 minutes should be allowed for these. We have not included draft answers so that, from the earliest stages of their careers, readers are encouraged to devise their own answers to problems arising in the field and the classroom.

Acknowledgements

The publishers are grateful to the following for permission to reproduce diagrams and photographs:

Airedale Air-Conditioning Ltd
Airserco Manufacturing Company
Alco Controls Division, Emerson Electric Company
Amana Refrigeration International, Inc.
ASHRAE (from 1977 Fundamentals, 1978 Applications, 1979 Equipment and 1980 Systems Handbooks)
Bally Case and Cooler Inc., Bally, Pa
Binder Engineering Co. Ltd
Carrier Corporation
Casella London Ltd
Danfoss
Delchi UK Ltd
Du Pont
D.W.M. Copeland
Eaton Williams Group Ltd
Electrolux Ltd
Hall Thermotank Products Ltd
Imperial Clevite Ltd
IRE
Jackstone Froster Ltd
Myson International Ltd
Prestcold Searle International Ltd
Pullen Pumps Ltd
Ranco Europe Ltd
R.A. Bennett Ltd
RCA Receivers Ltd
Sporlan Valve Company
Tecumseh Products Company
Trox Brothers Ltd
York, Borg Warner Ltd

We are grateful to the following Boards for permission to reproduce questions from past examinations papers:

City & Guilds of London Institute for questions from *Refrigeration 2* (Industrial Refrigeration) 2nd written, June 1977; *Refrigeration 1* Trade Studies 2, May 1979; *Refrigeration 2* (Air-Conditioning) Dec 1979.
The West African Examinations Council for q 1, 6 *Refrigeration Practice*, Theory (1) 1979; q 4, 5, 7 *Refrigeration Practice*, Theory (2) 1979; q 1, 3 *Refrigeration Practice*, Theory (2) 1980; q 1, 2, 4, 5 *Refrigeration and Air-Conditioning Practice*, Theory (2) 1981.

1. The history of refrigeration (WA)

Whilst it is only in recent years that refrigeration equipment has been mass-produced and widely used, its benefits will have been apparent to ancient man. Tribes living or hunting in regions close to the ice line will have noticed that the flesh of animals killed, or fish caught, during the winter kept well. But as the snow melted and summer came, food decayed more quickly. Closer to the equator high temperatures and humidity caused rapid putrefaction, so that thin strips of flesh dried in the sun became part of man's normal diet. They still are, in the form of pemmican and stockfish.

The value of ice as a preservative was known and put to use thousands of years ago. In winter it was cut, and moved into 'ice houses' which were built into the ground to make use of the insulating properties of soil. Such a structure was described by a Chinese poet, Shih Ching, about 3 000 years ago. In biblical times the Greeks and Romans harvested ice from the Alps for use in 'ice houses'. In the Middle East and India water was chilled by evaporating it through porous clay pots which were buried overnight – in favourable conditions it could be made cold enough to form ice. The Indian sub-continent also saw the first moves towards air-conditioning: water-soaked mats were hung across windows and doors to provide evaporative cooling during hot, dry weather. The Romans were more worried by chilly winds, and developed their underfloor central heating systems.

The pioneers

For some 1 500 years after this little progress was made and it was not until 1748 that the first moves towards the use of modern systems were recorded. In that year William Cullen investigated the effects of evaporating ethyl ether into a partial vacuum. But it was difficult to arrange good vacuums in those days, and it was not until 1834 that Jacob Perkins patented a closed-cycle refrigeration system using a compressor. In the same year he also patented a high pressure, hot water, heating circuit. Ten years later John Gorrie developed an air-cycle plant to make ice and to cool air for circulation through his hospital in Florida.

The race was really beginning now, with engineering innovations following quickly after new scientific theories. For man was moving into rapidly growing cities and huge cattle ranches were in prospect in the Americas and Australia; reliable, efficient refrigeration was essential if fresh meat was to be transported to the cities in sufficient quantity. The 1850s saw much development work on vapour compression systems. In the USA, Twinning's ice-making equipment earned renown, whilst James Harrison built units for meat freezing and brewery applications in Australia. Some years later the introduction of ammonia as a refrigerant enabled efficiency and reliability to be much improved. Ammonia inspired a new generation of reciprocating compressors and was used by Ferdinand Carré in the first viable absorption refrigeration system. Introduced in 1859, this used the ammonia-water cycle still used in absorption-type domestic refrigerators today.

That refrigeration came of age in the 19th century was due in no small part to the work of William Thompson, who later became Lord Kelvin. The products of his long life (1832–1907) reflect many interests ranging from refrigeration to telegraphy. Those which will interest us include the absolute temperature scale, and the law of conservation of energy. Not many years after him, Willis Carrier conducted the research which, when published in 1911, provided the foundations on which today's air-conditioning industry was built. The quality of this pioneering work, and its relevance to modern practices, can be illustrated by the reference to John Gorrie earlier in this chapter. Air-cycle machines may sound primitive, but they are still used in the air-conditioning systems of some jet aircraft!

In retrospect, refrigerants themselves did more than anything to hold up the development of refrigeration. Of those available many were poisonous, or explosive, or both, and others functioned only at pressures so high as to require compressors and system components to be constructed like battleships, with prime movers (driving motors or engines) to match. A refrigeration plant was of necessity bulky, and had to be constantly tended by well-trained and well-equipped operators. Small domestic refrigerators appeared only a distant hope, and daily deliveries of block ice to replenish 'ice boxes' were the only means of preserving foodstuffs at home. Air-conditioning was not a good prospect either: the most obvious applications for comfort cooling – cinemas and theatres – requiring safety standards which could not be reconciled with the circulation of poisonous and/or explosive refrigerants throughout the buildings.

Breakthrough

After having caused so many headaches for so long, the problems were solved with incredible speed as the 1920s drew to an end. In America the Vice President (Research) of General Motors – of which Frigidaire was then a division – became convinced one day in 1928 that refrigeration would get nowhere unless a new refrigerant could be found. The next morning this problem was passed on to research chemist Thomas Midgely. Midgely and his associates, Henne and McNary, had an idea that the answer would be a compound containing fluorine. It was, and it took them only three days to find it! they removed two atoms of chlorine from tetrachloromethane (carbon tetrachloride) molecules and substituted them with two atoms of fluorine to make the compound dichlorodifluoromethane, CCl_2F_2.

Tests confirmed that their discovery was a superb refrigerant, was non-flammable and had unusually low levels of toxicity. In 1930 two companies, General Motors and du Pont, formed a new company – Kinetic Chemicals Inc. – to manufacture and market it. Happily, neither salesmen nor customers were asked to follow the conventional practice of calling the refrigerant by its chemical name. Du Pont, who became sole owners of the new company in 1949, christened it Freon 12. It was the first of a quickly expanding group of fluorinated refrigerants used in the great majority of systems now manufactured, and is now made in many other countries and codenamed 'R12'.

The availability of R12 enabled equipment manufacturers to think in terms of smaller and more lightly built compressors and system components than had previously been necessary. New materials could be used – unlike ammonia, for example, R12 does not attack copper. Its non-toxic, non-explosive properties made it safe to use for air-conditioning as well as refrigeration applications. The engineers could rethink not only system designs, but manufacturing techniques, and in 1937 an engineering development which would, together with R12, literally bring refrigeration to the millions was unveiled by Tecumseh Products Co.: the first range of welded hermetic compressors.

The big boom

Developments were then restricted by World War II, but its end made manufacturing facilities available and there was a boom in refrigeration and air-conditioning which has continued up to the present. By the mid-1960s there were some 75 million refrigerators and food freezers in use in the USA alone, and several European manufacturers could count their annual production in terms of millions. And the boom in domestic appliances revolutionised the refrigeration industry and its commercial outlets. Bigger and better coldstores were needed at the docks certainly, but now coldstores were demanded by complete distribution chains: docks to main depots to district depots to butchers, fishmongers, grocers, bakers.... Refrigeration was not for imported foodstuffs alone. Demand came from all sides: plant to freeze vegetables close to the fields in which they had grown; plant to freeze fish on the trawlers which caught them; plant to freeze pre-cooked meals; poultry freezing factories; ice cream factories; factories to freeze pastry and cream cakes; more depots; refrigerated transport; refrigerated display cases ... and this was only for food. There were other areas for expansion: to answer medical needs for blood banks, for storing antibiotics, for making ice for compresses and low temperature surgery, for preserving samples of tissues that pathologists worked on. Industry used refrigeration in making antibiotics and cooling moulds in machines churning out such diverse products as candles and golf balls. The list could fill many pages.

The rapid expansion in refrigeration was accompanied by an almost equally dramatic increase in air-conditioning equipment sales to cinemas and theatres, office buildings and hotels, hospitals and airport terminal buildings, departmental stores and banks and shops, large and small. A demand grew up for household goods: 'window' units and small packaged plant and millions of heat pumps. The new methods were used widely in industry: textile factories and printing plants, the areas handling the high quality steel cords for radial tyres, precision engineering plants and the pharmaceuticals companies making anhydrous powders in the most humid parts of Africa among others. New applications are constantly being added by the increased applications of electronic equipment; computer suites, telecommunications installations, avionic service areas and factories for the manufacture of microchips themselves are typical of places in which precise control of temperature and humidity is essential. The only factor which seems likely to reduce demand is the cost and availability of fossil fuels. But the heat pump can already produce much more thermal energy than it consumes; and one wonders what new arrangements of atoms the chemists are now developing, and how far the engineers can reduce the energy needs of new equipment. They have not failed us so far!

2. Heat, temperature and pressure

To understand the theory of refrigeration one first needs to examine some concepts used in physics. Among the most important for us are *heat* and *temperature*.

Heat

Heat is a form of energy and is measured in joules (calories in the metric system or Btu in the British system). The heat of a body is related to the movement of the atoms and molecules that make it up. When a body is very hot its molecules and atoms move very rapidly and over relatively long distances. As it is cooled (reducing heat energy) its molecules and atoms move less vigorously until the atoms are vibrating quite gently. When molecules or atoms are moving relatively slowly the substance is liquid. If it loses heat energy they move more slowly until the substance reaches its *freezing point*.

The freezing point is that temperature at which the removal of heat will cause the liquid to solidify. (Imagine the molecules clinging together, to keep each other warm.)

If, on the other hand, the liquid is heated its molecules (or atoms) will move more quickly until the *boiling point* is reached.

The boiling point is that temperature at which the addition of any heat will cause a liquid to turn into vapour. (The ever-moving particles now resemble aircraft tearing along a runway. They have reached take-off speed but have not yet lifted off.)

We cannot measure the movement of molecules directly, nor can we take all the heat out of a body to see how much heat there is in it. So when we want to know how hot something is we do not measure its heat content, but we measure its temperature using a thermometer.

Units of heat

In the SI system of units heat energy is measured in joules, but normally expressed in terms of kilojoules (1 kJ = 1 000 J) or megajoules (1 M = 1 000 000 J). It is sometimes expressed in watt-hours or kilowatt-hours (1 kWh = 3·6 MJ).

The British thermal unit (Btu)
One British thermal unit is the amount of heat required to raise the temperature of 1 lb of water by 1 °F.

The calorie (cal)
One calorie is the amount of heat required to raise the temperature of one gram of water by 1 °C.

Conversion factors for the various units of measure are given in Chapter 14, but as an approximate comparison 1 Btu equals about 250 cals. For our purposes the calorie is an inconveniently small quantity of heat and heat is often expressed in terms of kilocalories (kcal) which is equal to approximately 4 Btu.

Unlike these measures the joule is not defined in terms of temperature rises in water, but in terms of other physical quantities and it is found that the heat energy required to raise 1 kg of water through 1 °C is 4 200 J, i.e. 1 kcal = 4 200 J.

Temperature

The temperature of a substance establishes its position on a scale of thermal activity.

The scale used to measure temperature is chosen for its convenience in making calculations and at different times different scales have been chosen. In all of them *fixed points* are chosen and the positions of the other points are worked out from these. You will come across various scales and you should be familiar with the following.

Celsius (or centigrade)
There are two fixed points on this scale – the boiling point and the freezing point of pure water at sea level. The freezing point (lower fixed point) is given a value of 0 °C and the boiling point (upper fixed point) is

HEAT, TEMPERATURE AND PRESSURE

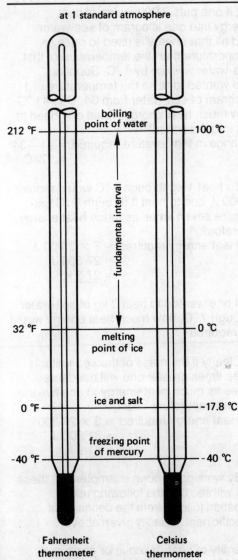

Fig. 2.1 Fahrenheit and celsius thermometers

given a value of 100 °C. Between the freezing and boiling points the scale is divided into 100 equal divisions each of 1 °C.

Fahrenheit
The scale gives different values to the fixed points mentioned above. The freezing point of water is at 32 °F and the boiling point of water is at 212 °F. The scale between these points is divided into 180 equal divisions.

The arithmetical relationship between the celsius and fahrenheit scales is shown in Fig. 2.2.

Kelvin
This scale (named after a famous physicist) has a fixed point at absolute zero. It was mentioned above that the movement of molecules or atoms in a substance depends on its heat content. As the heat content falls the movement gets less until, at absolute zero, the movement stops altogether. At this point the substance has no heat content. In practice, absolute zero has never been reached, although some very low temperatures very close to it have been achieved in experiments. Its value has been calculated to be −273·1 °C (−459·6 °F).

On the Kelvin scale absolute zero has a value of 0 K. The scale has divisions equal in size to those of the celsius scale so that 0 °C is equivalent to 273·1 K, 100 °C to 373·1 K, 26 °C to 299·1 K, etc.

Rankine
You may also see references to another scale, Rankine. It also has a fixed point at

---

To convert a temperature given in fahrenheit to celsius.

Temperature on the celsius scale = Temperature on the fahrenheit scale − 32 × $\frac{5}{9}$

e.g. To express 50 °F in celsius

Temperature in °C = $(50 - 32) \times \frac{5}{9} = 18 \times \frac{5}{9} = 10$ °C

To convert a temperature given in celsius to fahrenheit.

Temperature on the fahrenheit scale = Temperature on the celsius scale × $\frac{9}{5}$ + 32

e.g. To express 5 °C in fahrenheit.

Temperature in °F = $\left(5 \times \frac{9}{5}\right) + 32 = 9 + 32 = 41$ °F

Fig. 2.2 Fahrenheit and celsius conversion examples

TROPICAL REFRIGERATION AND AIR-CONDITIONING

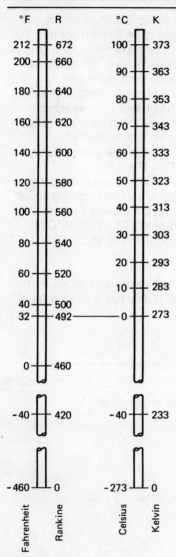

Fig. 2.3 Absolute temperature conversions

absolute zero, but has degrees of the same size as those on the fahrenheit scale so that 0 °F is equivalent to 459·6 R, the freezing point of pure water is equivalent to 491·6 R and the boiling point of pure water to 671·6 R.

Specific heat capacity

It was mentioned at the beginning of this chapter that when one measures temperature one is not measuring the heat content of a body. One of the reasons for this is that when the heat content of a body increases by a given amount, the temperature increase will depend on the material of which the body is made up. If one takes a kilogram of iron and a kilogram of water and puts the same amount of heat energy into each, the temperature change in the iron will be about nine times greater than in the water. It is clearly important to know how much heat one is going to need to put into a substance in order to make it reach a certain temperature and to help us with this we use *specific heat capacities*.

The specific heat capacity of a substance is the quantity of heat required to raise the temperature of 1 kilogram of that substance through 1 °C (or in the British system 1 pound through 1 °F).

In the SI system specific heat is expressed in kJ/kg°C. For example, the average specific heat of pure water is 4·2 kJ/kg°C. In the British system, specific heat is expressed in Btu/lb°F and the specific heat of pure water is 1·00 Btu/lb°F.

Let us study an example. The specific heat capacity of sea-water is 3 900 J/kg°C; i.e. if one puts 3 900 joules of heat energy into one kilogram of sea-water and all that energy is used to raise its temperature then the temperature of that sea-water will rise by 1 °C. Suppose one wanted to raise the temperature of 1 kilogram of sea-water from 34 °C to 41 °C, how much heat energy would one need to put into the water?

Change in temperature required = 41 − 34
$$= 7 \ °C$$

To heat 1 kg through 1 °C would require 3 900 J, and to heat it through 7 °C will require seven times as much heat energy, therefore:

Heat energy required = 7 × 3 900 J
= 27 300 J
= <u>27·3 kJ</u>

If one wanted to heat 3 kg of sea-water through 7 °C, how much heat energy would be required?

Clearly if the mass of the sea-water is three times greater one will need three times as much heat energy to achieve the same temperature change.

Heat energy required = 3 × 27 300
= 81 900 J
= <u>81·9 kJ</u>

By working through examples like these you will see that the following useful equation follows from the definition of specific heat capacity given above.

Quantity of heat (taken in or given out)
= Mass × Specific Heat Capacity
× Change in Temperature

The above formula is therefore used in all calculations affecting cooling or heating processes, to help determine the required equipment capacity.

Sensible and latent heat

When the heat content of a body changes, two things can happen: either the temperature of the body can change, or if the body is at its melting or boiling point, it can change state.

A change in the sensible heat of a body results in a change in the temperature of that body. For example, if one heats a beaker of water that is at 20 °C (68 °F) its temperature will rise as its sensible heat is increased.

The addition or subtraction of latent heat to or from a body results in a change of state WITHOUT A CHANGE IN TEMPERATURE. For example, if one is boiling water in a kettle and the temperature of the water reaches 100 °C (212 °F) and one continues to heat the kettle further, some of the water is changed into steam which escapes from the spout. The heat energy being put into the water is changing the state of the water, but the temperature of the steam does not rise above 100 °C (212 °F) while there is still water in the kettle.

For nearly all substances there are two values for latent heat, not just one.

a) The latent heat of vaporisation (which has the same value as the latent heat of condensation) is the quantity of heat required to change a unit mass of a liquid into the gaseous state (or given out when a unit mass of gas changes to the liquid state) without a change in temperature.
b) The latent heat of fusion is the quantity of heat required to change a unit mass of a solid into the liquid state (or given out when a unit mass of a liquid changes into the solid state) without a change in temperature. To give you an idea of the quantities of heat involved here are the values for water.

Latent heat of vaporisation of water
= 2 260 kJ/kg (970 Btu/b)
Latent heat of fusion of water
= 336 kJ/kg (144 Btu/b)

Comparing these latent heat values with the energy needed to change the temperature of pure water at sea level, we see that the latent heat of vaporisation is more than five times the quantity of heat required to raise the temperature of water from freezing point to boiling point. The possibility of being able to make use of such a large quantity of heat while only causing minor changes in sensible temperatures must have interested pioneers designing the first refrigeration systems. The facts and figures are important, so they are summarised in Fig. 2.4.

Heat transfer

The heat content of a body can only increase if it receives energy from outside or if it converts another form of energy into heat energy. Heat energy cannot be created out of nothing because, as stated in the Law of Conservation of Energy: **Energy can neither be created, nor destroyed.**

Similarly, when a body loses heat, that heat must be transferred to something else either as heat energy or another form of energy. Examples of energy being converted from one form into another surround us in our everyday lives: light energy from the sun being converted into chemical energy in plants, electrical energy being converted to mechanical energy in electric motors and so on. Energy can also be transferred in the same form from one substance, or space, to another and this is basically what refrigeration is all about. **Refrigeration is the process of removing heat from one space or substance, and transferring it to another.**

All appliances used for refrigeration rely on the latent heat of vaporisation of various substances. If water evaporates inside a heat exchanger in an insulated cabinet, it will absorb heat from the cabinet and it's contents. If the water vapour is then drawn out of the heat exchanger and cooled so that it condenses back into a liquid, the heat picked up inside the cabinet will be rejected into the surrounding air.

How, and why, can heat be transferred from one medium to another? There are three methods:

a) **Conduction is the transfer of heat energy from a body at a higher temperature to a body at a lower**

TROPICAL REFRIGERATION AND AIR-CONDITIONING

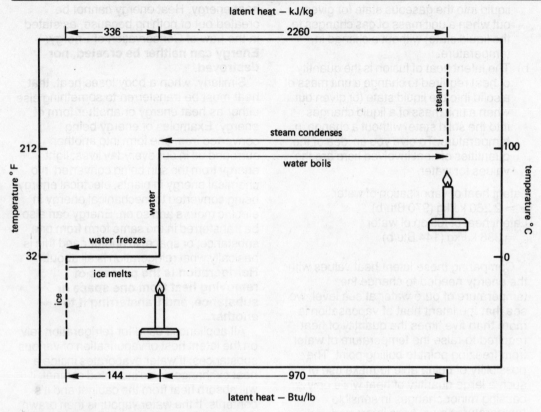

Fig. 2.4 Sensible and latent heat relationships

water over a flame, the water at the bottom will expand and become less dense while that at the top will remain cool and not expand. It will sink allowing the less dense water to rise to take its place. This water, now at the bottom, is in turn heated and the warmer water above it will be cooled by the air in contact with its surface, and the sides of the jar. It will fall as freshly-heated water rises, and a circulating pattern of convection currents will be established.

Fig. 2.5 Convection currents

temperature through direct contact. Here, the atomic and molecular activity mentioned in the first paragraph of this chapter has a 'knock-on' or 'shunting' effect. Hotter, faster-moving particles collide with and transmit part of their energy to cooler, slower-moving ones next to them. These repeat the process until, eventually, temperature levels equalise. The process can take place in solids, liquids, or gases; or at the boundaries between materials. Heat one end of a rod of copper, and you can feel it happening.

b) **Convection is heat transfer resulting from the movement of a liquid or a gas.** A warm liquid or gas will rise; a cool one will fall. If we heat a jar of

HEAT, TEMPERATURE AND PRESSURE

c) **Radiation is the transfer of heat from one substance to another over a gap which separates them.** A relatively intense source of heat emits waves similar to light waves or radio waves. These waves are received by cooler matter, and converted back into heat energy.

It is essential to remember that heat will always flow from a body at a higher temperature to one at a lower temperature. It resembles water – always flowing downhill, and trying to attain an equal level. Air behaves in a similar way – flowing from warmer to cooler areas, from higher to lower pressure zones, always trying to equalise pressure. If we break a barrier between air at atmospheric pressure and a container from which most of the air has been evacuated, the higher pressure air will rush into the vacuum, until the pressures reach an equal level.

Pressure

In defining units of thermal measure, and describing thermometers, we have said that the water used to set the scale must be pure and at sea level. The first point is obvious, since impurities would alter the physical properties of the water including its freezing and boiling points. The second is more complex, and has more to do with air pressure than height itself. The temperatures at which changes of state occur depend on pressure, and to set temperature scales we use atmospheric pressure.

To the refrigeration engineer, pressure is as important as temperature. We must understand the causes of pressure, the instruments with which it is measured and the units of measure before we can clearly understand the refrigeration cycle. Pressure is force per unit area. The unit in the SI system is a pascal (Pa), 1 Pa being the same as 1 N/m^2 or 10^{-5} Bar. A more convenient measure is a kPa, which is $1\,000 \text{ N/m}^2$. In the British system pressure is measured in pounds force per square inch, for which the abbreviations lbf/in^2 or psi can be used.

Atmospheric pressure

Atmospheric pressure is the weight of air above us per unit area. Since the value of atmospheric pressure measured from day to day varies slightly we have a standard called a *standard atmosphere*, which can be used for test purposes and when defining other units of measure. This pressure is specified as 101·325 kPa ($14·7 \text{ lbf/in}^2$).

Atmospheric pressure can be measured by a mercury barometer. Such a barometer can be made from a glass tube approximately 1 m (3 ft) long, with one end sealed. The tube is filled with mercury, and the open end is inserted below the surface level of more mercury in an open bowl. The mercury in the tube does not 'collapse' into the bowl. Atmospheric pressure pushing down on the mercury in the bowl will maintain the top of the column in the tube approximately 760 mm (30 in) above the surface level in the bowl, at which point the pressure of the atmosphere is balanced by the weight of the mercury column acting on the area of the column. This height will change in accordance with variations in atmospheric pressure and can easily be measured if the tube is calibrated. Using the symbol for mercury, Hg – a column of 760 mm (30 in) Hg is defined as to one standard atmosphere; 760 mm (30 in) Hg is therefore equivalent to a pressure of 101·325 kPa ($14·7 \text{ lbf/in}^2$).

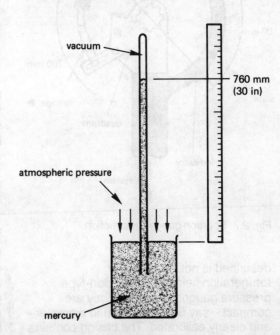

Fig. 2.6 Mercury barometer

Pressure gauges

In refrigeration we need to measure, not atmospheric pressure, but the pressures within containers of various kinds. For this the mercury barometer in the form

TROPICAL REFRIGERATION AND AIR-CONDITIONING

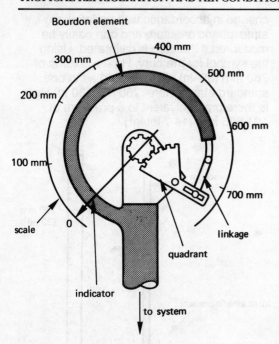

Fig. 2.7 Bourdon gauge construction

described is not suitable. For refrigeration field work, Bourdon-type pressure gauges are used. They are compact – say 60 mm (2·5 in) in diameter – and clearly calibrated. The casing contains a bronze or steel tube of oval cross-section curled into a question mark. The inner end of the tube is sealed, and the other open to the pressure to be measured. Variations in pressure cause the tube to 'flex' from its rest position; and this movement is transmitted through a quadrant and mechanical linkage to a pointer. This indicates, on a calibrated scale, the degree of pressure – or vacuum – to which the tube has been subjected, *in terms of its difference from atmospheric pressure.* Exposed to the pressure of a standard atmosphere, a pressure gauge would read zero. This type of instrument is not accurate when reading vacuums of more than 28 in Hg (710 mm Hg) on its scale, i.e. vacuums with a pressure of less than 50 mm Hg (2 in Hg) in absolute terms. **Gauge pressure is that by which the pressure of a system exceeds or falls short of atmospheric pressure. Absolute pressure is the sum of gauge pressure and atmospheric pressure, or, in the case of a partial vacuum, atmospheric pressure minus gauge pressure.**

Be careful not to confuse gauge and absolute pressures, and to check that any data sheets you are using clearly define which of the two they are quoting.

Vacuum

Our reference to *vacuum* calls for another definition and another unit of measure. A perfect vacuum is the total absence of all pressure. This may not exist, even in outer space, but if it does its absolute pressure reading will be zero. As a gauge pressure in terms of the height of mercury it is 760 mm (30·0 in) Hg of vacuum.

The expression of vacuum as a height of mercury can be confusing so we'll explain it. Imagine a long tube open at both ends, standing vertically in a bowl of mercury. At this stage the mercury inside and outside the tube are at the same level. The exposed end is now connected to a vacuum pump and a partial vacuum is created in the tube. Some mercury rises up the tube to a height – say – of 254 mm (10 in) – we say that the pump has 'pulled a vacuum of 254 mm'. The pressure on the surface of the mercury in the bowl is balanced by the pressure caused by the mercury column *plus* the pressure caused by the air remaining in the tube.

Atmospheric pressure =
  pressure due to mercury column
  + pressure of air on column

Now, atmospheric pressure is equivalent to 760 mm (30 in) Hg, therefore:

760 mm (30 in) =
  254 mm (10 in)
  + pressure of air on column

therefore: pressure of air on column is 506 mm (20 in) Hg, which is equivalent to 67·5 kPa (9·8 lbf/in). We would say that the air in the tube has a *gauge pressure* of 254 mm (10 in) Hg of vacuum (or 34 kPa, 4·9 psig), but its absolute pressure is 67·5 kPa (9·8 psi).

More accurate measurements of deep vacuum than are possible with the Bourdon-type gauge can be made using a *manometer*, or 'U' tube. One type of manometer often used in refrigeration work is shown in Fig. 2.8(a). It will be seen that if the open leg is connected to a source of partial vacuum, the mercury will rise up it; whilst the level in the other leg falls. If leg A is connected to a perfect vacuum the tops of the mercury columns in both legs will read zero. As the pressure in leg A rises, and the mercury in leg A falls the user must

HEAT, TEMPERATURE AND PRESSURE

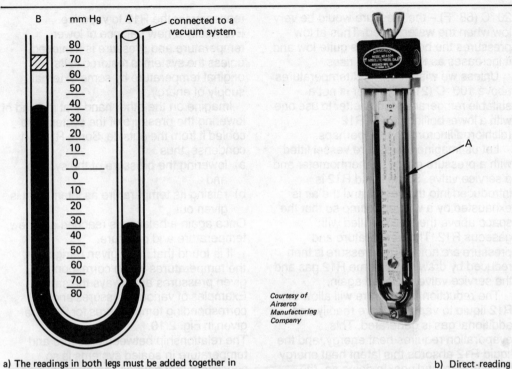

a) The readings in both legs must be added together in this manometer

b) Direct-reading manometer

Courtesy of Airserco Manufacturing Company

Fig. 2.8 Manometers

add together the readings for leg A and leg B in order to find the value of the pressure being measured. This manometer can read absolute pressures of up to 125 mm Hg.

To reduce the risk of error, the above scale is often replaced by a direct-reading scale calibrated in either British or metric units (see Fig. 2.8(b)). In this manometer inches of vacuum are read by noting the level of the mercury in leg A only. Absolute Pressure of the vacuum can be calculated to a degree of accuracy approaching 1 mm Hg of pressure.

But even the millimetre is not a fine enough unit to measure, or express, a nearly perfect vacuum. And it is necessary to do so, since, for absolutely efficient and reliable dehydration, refrigeration systems need to be evacuated to absolute pressures of even less than 1 mm. For this level of accuracy we need to use an electronic high vacuum gauge, which will register pressures of the order of 20 thousandths of 1 mm Hg. That is itself a clumsy expression, and the pressure is referred to in terms of Microns (a micron is one thousandth of 1 millimetre). Thus the pressure gauge will register pressures of 20 microns Hg (20 μm Hg).

Pressure/temperature relationships

If the boiling point of water was tested in an unpressurised aircraft and deep underground in a gold mine where air pressure is high, it would be seen that changes in atmospheric pressure cause changes in the boiling point. At 3 000 m (10 000 feet) the air pressure would be approximately 69 kPa (10 lbf/in^2) and water would boil at 90 °C (193 °F). At 10 000 m (30 000 feet) air pressure would be only 34 kPa (5 lbf/in^2) and the boiling point down to 73 °C (163 °F). But in the mine, things would be very different. At 2 600 m (8 000 feet) below sea level the air pressure would have risen to 138 kPa (20·0 lbf/in^2) and water would not boil until its temperature reached 109 °C (228 °F). There is a clear relationship between pressure and boiling point.

Sealed system pressure relationships

These facts could be of help in designing a simple refrigeration system, but have one major shortcoming. They relate to processes affected by atmospheric pressure, and the refrigeration engineer cannot afford to have performance change with the weather, or see his refrigerant continually vanishing 'into thin air'. He must use a sealed circuit, and we must see what effect this will have on the pressure/temperature relationship.

A simple experiment illustrated in Fig. 2.9 can show that boiling point varies with pressure in a sealed system as well as in the atmosphere. A stoppered flask containing some water (below 100 °C) and air is connected to a vacuum pump. When the pump is switched on the pressure in the flask falls gradually until the water is seen to boil. If the temperature of the water is – say – 90 °C (194 °F) the pressure will not have to fall much before the water boils, but if the temperature is lower – say 20 °C (68 °F) – the pressure would be very low when the water boiled. Thus at low pressures the boiling point is quite low and it increases as the pressure rises.

Unless we wish to work at temperatures above 100 °C (212 °F) water is not a suitable refrigerant. Much better to use one with a lower boiling point – R12 (dichlorodifluoromethane) perhaps.

Let us imagine a pressure vessel fitted with a pressure gauge, a thermometer and a service valve. Some liquid R12 is introduced into the vessel and the air is exhausted by a vacuum pump so that the space above the liquid is filled with gaseous R12. The temperature and pressure are noted. The pressure is then reduced by drawing off some R12 gas and the service valve is closed again.

The reduction in pressure will allow the R12 liquid to vaporise more readily and additional gas is generated. This evaporation requires heat energy, and the liquid R12 absorbs this latent heat energy from its surroundings. In doing so, *the temperature of the system is reduced* to a lower level which, in turn, reduces the tendency for the R12 to vaporise. Eventually a new balance of lower temperature and pressure is achieved, unless the system is restored to its original temperature by some external supply of energy.

Imagine on the other hand that instead of lowering the pressure of the system one cooled it from the outside. Some R12 will condense, thus
a) lowering the pressure of the system, and
b) raising its temperature as latent heat is given out.

Once again a balance is reached at a new temperature and pressure.

It is found that, for a given refrigerant, the temperatures which correspond to given pressures are always the same. Examples of various pressures and the corresponding temperatures for R12 are given in Fig. 2.10.

The relationship between pressure and temperature in sealed systems is so consistent that gauges can be calibrated in terms of the temperature of a given refrigerant, as well as its pressure.

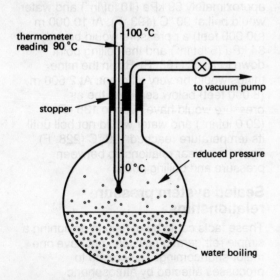

Fig. 2.9 Apparatus to show the variation of boiling point with pressure

Pressures			Temperatures	
kPa	lbf/in²	kgf/cm²	°C	°F
322·0	46·7	3·283	10	50
397·8	57·7	4·056	15·5	60
521·3	75·6	5·315	23·3	74
643·4	93·3	6·559	30	86

Fig. 2.10 R12 pressure/temperature examples

Theoretical refrigeration cycle

This additional information enables us to design a simple mechanical refrigeration system, and describe its cycle of operation. Follow the details of the cycle by looking at Fig. 2.11.

Using a hermetically sealed circuit, *high pressure* liquid is metered through a flow-control valve into a cooling coil which is maintained at a *low pressure*, and installed inside an insulated cabinet. Exposed to this low pressure, the liquid refrigerant boils. The latent heat necessary to secure the change of state is transmitted through the walls of the cooling coil from the cabinet and its contents, which are at a higher temperature than the refrigerant vapour. The vapour, now warmed to a relatively high temperature, is drawn into to the compressor. This compresses the vapour, and discharges it as a *high pressure* gas close to the pressure at which it will condense. The vapour is pumped through a second cooling coil located in cool ambient air outside the cabinet. Heat is transmitted from the hot vapour to the cooler air, so that the refrigerant condenses back into its liquid state. Still under high pressure, the liquid refrigerant flows back to the control valve, and the cycle is repeated.

If that simple description is rewritten, it can be used to introduce and explain several technical expressions, with which we must be completely familiar.

High pressure liquid refrigerant passes through a *refrigerant control* (which could, incidentally, be either an *expansion valve* or a *capillary tube*) into the *evaporator* inside the refrigerator. This is maintained at a low *suction pressure* and the liquid *evaporates* at the temperature which corresponds to that pressure. In doing so, it collects heat from the cabinet and its contents, and the gaseous refrigerant is *superheated*. After being recompressed, the refrigerant – now a *hot gas* in the form of *saturated vapour* – is pumped into an *air-cooled condenser*. Heat from the gas is

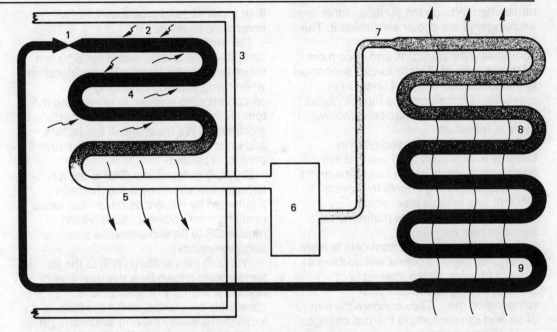

1 Refrigerant control　　2 Cooling coil (evaporator)　　3 Insulated cabinet　　4 Refrigerant 'boils' and absorbs heat
5 Refrigerant superheated on leaving the evaporator　　6 Compressor
7 High pressure hot gas discharges into the condenser　　8 Heat is lost to ambient air and refrigerant condenses
9 Refrigerant leaves the condenser as a sub-cooled liquid

NB Evaporator and suction line are called 'low side' items and the discharge line, condenser and liquid line are called 'high side'.

Fig. 2.11 Theoretical mechanical refrigeration cycle

transmitted through the surface of this *heat exchanger* to the cooler air outside it. The saturated vapour is cooled to its *condensing temperature*, and once more becomes a liquid. This is further *subcooled* in the condenser and, still under high pressure, is forced back to the refrigerant control before the mechanical *refrigeration cycle* is repeated.

These terms will be used often in Chapter 5, where we shall see that this basic circuit needs only a few refinements and additional components to provide efficient and reliable mechanical refrigeration for domestic refrigerators, food freezers etc.

Before studying equipment details more closely, we need to note a few additional rules of physics, which interest both air-conditioning specialists and refrigeration men. They concern the nature of air and its reactions to thermal changes.

The properties of air

Air is a mixture of many gases – principally nitrogen and oxygen – and water vapour. At sea level, the pressure of the air forming the earth's atmosphere is 101 325 Pa (14·7 lbf/in^2) and its density – its mass per unit volume – is approximately 1·2 kg/m^3 (0·074 lb/ft^3) at 21 °C (70 °F).

As previously noted, air pressure decreases with increased altitude and it becomes less dense. Atmospheric pressure changes with height and is affected not only by variations in temperature, but also by variations in the air's moisture content. Dry air is denser than moist air, and produces a higher barometric pressure.

The amount of moisture present in the air both inside and outside a building is a major factor to be taken into consideration when designing and balancing an air-conditioning system, or preventing the formation of condensation on the surface of a coldstore. We must define the factors and units of measure to which moisture content is related!

Dry bulb temperature (DB) is the air temperature shown by a thermometer unaffected by influences other than those resulting from solar exposure (which require DB to be expressed as *shade* or *sunlit* measures).

Wet bulb temperature (WB) is the air temperature shown by a thermometer of which the mercury or spirit bulb is constantly moistened, whilst the bulb is located in a swiftly moving airstream. (In practice, the wet bulb is itself normally rotated as quickly as possible in a relatively still air mass.) In a sling psychrometer one dry and one wet bulb thermometer are rotated to measure DB and WB temperatures simultaneously.

The DB is invariably the higher of the two readings, the difference between them being called the *wet bulb depression*. This increases as the volume of water vapour in the air decreases, and gives a clear indication of *relative humidity*. This is the quantity of water vapour present in relation to the maximum quantity which the air is capable of holding at a given dry bulb temperature. This quantity is related to the air's dew point temperature – the temperature at which water vapour would start to condense onto a cooled surface within the air mass. Water vapour exerts a pressure on the air which holds it, and air's ability to withstand that pressure is reduced if its temperature falls. If dew point is reached the vapour pressure 'wins' and some of the water vapour escapes and forms condensation.

Where one has two different gases and not two forms of the same substance, one has to take note of Dalton's law of partial pressure.

Dalton's law of partial pressure states that each constituent of a mixture of gases behaves as if it alone occupies a given space and that total pressure of the mixture of gases equals the sum of the pressures that would be exerted by each constituent if it alone occupied the space occupied by the mixture. Thus 1 g of water vapour, if occupying a space of 1 litre will have the same pressure whether it is mixed with air or not.

When moist air is cooled to a temperature at which part of its moisture content starts to condense – its saturation temperature – it will have lost sensible heat from both the dry air and the moisture, and latent heat from the moisture which has condensed. The sum of the two is the change in *enthalpy*. Enthalpy is the total heat content and, in the case of air, this is the heat content of the dry gases plus that of the water vapour which it contains. Enthalpy is expressed in terms of heat content per unit mass – J/kg (Btu/lb).

All of these varying measures can be expressed on a single chart, which is of great practical value to all of us, despite the unfortunate tendency to regard it as

HEAT, TEMPERATURE AND PRESSURE

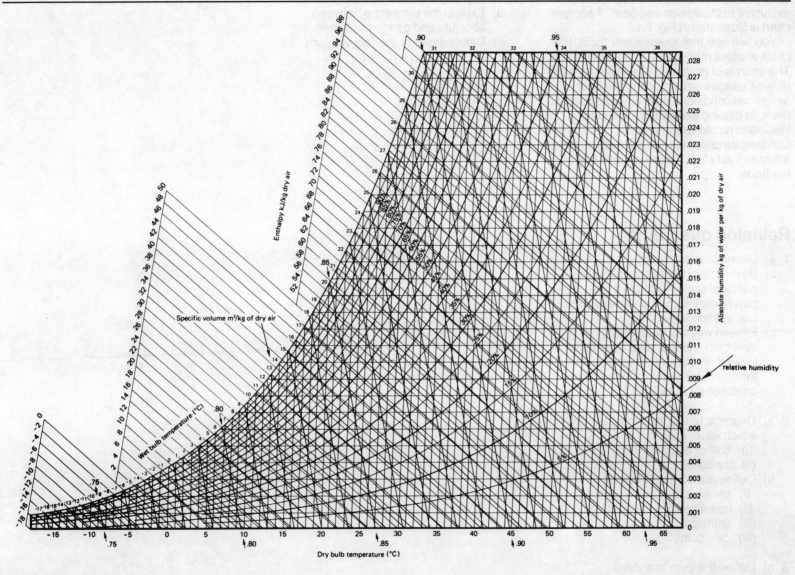

Fig. 2.12 Psychometric chart

exclusive to the design engineer. A sample chart is illustrated in Fig. 2.12.

You will see that given a few facts, the chart enables many more to be determined. The chart is of great practical value for such diverse usages as calculating the best air-off-coil conditions in an air-conditioning plant, to helping decide the thickness of insulation necessary to prevent condensate dripping from the suction line of a frozen food store. It is a valuable tool of the trade.

Revision questions

1. a) Define psychrometry.
   b) By means of a diagram, show the principal scales found in a typical psychometric chart and explain the use of *each* scale.
   c) Explain why the reading of a 'wet-bulb' thermometer is usually lower than that of a 'dry-bulb' thermometer in the same ambient conditions. (WAEC)

2. a) Describe briefly the processes by which water may:
      (i) 'boil' or evaporate,
      (ii) condense.
   b) Define each of the following terms:
      (i) sensible heat,
      (ii) latent heat,
      (iii) enthalpy,
      (iv) dew point.

3. a) Define the term 'standard atmospheric pressure'.
   b) Explain the difference between absolute and gauge pressures.
   c) Define the term 'perfect vacuum'.

3. Electricity – theory (WA)

In the last chapter we said that heat resulted from the movement of atoms and molecules. Electricity results from the movement, not of whole atoms, but of subatomic particles. To understand this it worth spending a few paragraphs of our time examining theories not required by the syllabus.

Atomic structure

Molecules are made of atoms and, minute though they may be, atoms are themselves made up of very much smaller particles. The most convenient atom to look at first is the most simple – hydrogen. Greatly magnified it would look like a moon in orbit round a planet as shown in Fig. 3.1. The 'planet', or nucleus, is a single proton in the case of hydrogen, which has a positive electrical charge. The 'moon' is an electron, which carries a negative charge. The two charges exactly balance each other, and the complete atom is electrically neutral.

More complicated atoms can contain many electrons, all in constant orbit round a nucleus which will contain a compact group of protons carrying a positive charge. The number of protons equals the number of electrons and their charges balance each other.

Electrons are not scattered at random. They are in layers, like the skins of an onion. Each layer has a fixed maximum possible number of electrons – up to two in the innermost layer, eight in the second layer, eighteen in the third, and so on. In some atoms the electrons are so arranged that the outermost electrons are not easily removed and the substances made up of such atoms will conduct little or no electricity and will, in electrical terms, be an *insulator*.

In certain other atoms the electrons are so arranged that they are easily removed and the atomic structure offers little resistance to the flow of electricity. It is in fact a *conductor*. Materials which lie between the two in terms of ability to conduct electricity are called dielectrics. Copper is an excellent conductor; it has electrons which are only loosely bound to the remainder of the atom and which can therefore carry electricity.

The electron theory

In conductors electrons are constantly being detached from atoms and they move about in a random manner as illustrated in Fig. 3.2.

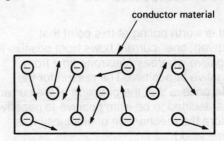

Fig. 3.2 Random movement of electrons

If an attractive force – in the form of an electrical charge – is applied to the conductor, the electrons will flow towards it, away from a repellent charge as shown in Fig. 3.3. This directed flow of electrons is called an electric current. As the positive pole of the battery attracts and draws off electrons from the conductor material, fresh electrons are provided by the negative pole of the battery so that the conductor material is always electrically neutral.

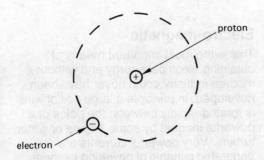

Fig. 3.1 Hydrogen atom

TROPICAL REFRIGERATION AND AIR-CONDITIONING

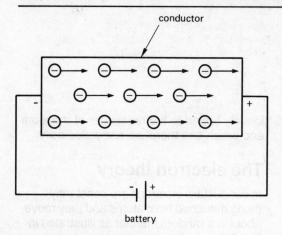

Fig. 3.3 Directed electron flow

It is worth noting at this point that 'conventional' current flows from positive to negative whereas electrons flow from negative to positive. The reason for the difference is that the direction of the current was decided to be from positive to negative before the mechanism of electricity was understood.

How currents are generated

Just as water remains still until an applied force makes it move, so the electrons in a conductor must be subjected to electrical 'pressure' or potential to make them move in one direction and cause a current to flow. In the example given above a battery was used to provide the energy required to cause the current to flow. There are several ways in which loose electrons can be 'shunted' into motion and a current generated.

Electro-chemical

This is a familiar way of generating a current, with dry cells used in torches serving as a practical, everyday example. Consider the torch battery shown in Fig. 3.4. When the terminals are connected into a circuit, electrons will be drawn from the zinc outer casing to the positive carbon rod at the centre of the battery, through the circuit. Inside the battery the electrons are removed from the carbon rod by a series of chemical reactions so that a negative charge does not build up on it and, similarly, chemical reactions ensure that a positive charge does not build up on the zinc plate.

These dry cells are called primary cells and the chemical reactions are non-reversible. Once the chemicals are used up they have to be thrown away. Accumulators, or secondary cells, have a reversible chemical reaction and can be recharged and re-used time and time again. They are used in motor cars to start the engine and work the lights.

Heat

When a source of heat energy is applied to a pair of wires made from dissimilar materials which have been twisted or welded together a very small current is generated. As sketched in Fig. 3.5 it seems too crude to be true, but in a more sophisticated form, it commands a good price when sold as a thermocouple.

Light

When light falls on a photoelectric cell a very small current is generated. This effect is used in photographers' light meters.

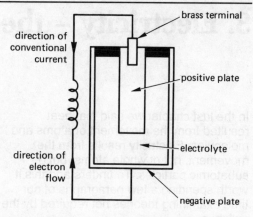

Fig. 3.4 Cross-section of a torch battery

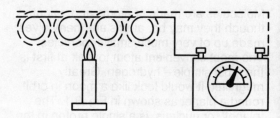

Fig. 3.5 Crude thermocouple

Electro-magnetic

This is the most important means of obtaining electrical energy and without it modern industry could never have been developed. In principle a large coil of wire is rotated rapidly between the poles of a powerful magnet by some engine or water turbine. Very powerful currents are generated capable of powering factories and lighting whole towns. The power output of a modern generating station using electro-magnetic generators can exceed two thousand million watts of power continuously compared with a few watts of power from a dry battery.

Basic principles of current electricity

Let us consider a simple circuit in which a cell is connected by wires to a light bulb. When the circuit is complete the bulb lights up showing that a *current* is flowing.

Current

The current is the rate of flow of electricity through a circuit and is measured in amperes (A). One ampere is said to be flowing when 6.3×10^{15} electrons pass any point in the circuit in one second. The current in a series circuit must be the same at all points (otherwise there would be a build-up of electrons at one point and this cannot happen). Current can be measured using an *ammeter*.

When a current flows, electrons are moving around the circuit, the number of electrons that flow depending on the strength of the current and the time for which it flows. The quantity of electricity that flows is measured in coulombs. One coulomb of electricity flows past a point in a circuit when 1 ampere flows in that circuit for 1 second.

Electromotive force

The cell has an *electromotive force* (e.m.f.) which drives the current round the circuit.

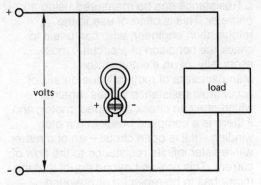

Fig. 3.6 Voltmeter connections

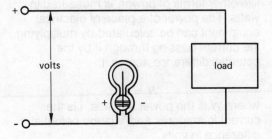

Fig. 3.7 Ammeter connections

This force is measured in volts. This e.m.f. sets up *potential differences* (p.d.) across various components in a circuit (in this case the biggest p.d.s are across its own terminals and across the light bulb). The units for potential difference are also volts and voltages can be measured using a volmeter. The voltmeter is connected across a component (the positive side of the circuit being connected to the positive terminal of the meter, and negative to

negative) to find the potential difference across it. You might expect the potential difference across the terminals of the cell when a current is flowing to be equal to the e.m.f. of the cell. This is *not* the case. The e.m.f. is the potential difference across the cell when *no* current is flowing. When a current flows the potential difference across the cell is less than the e.m.f. because some potential difference is required to drive the current through the cell itself.

Resistance

Any circuit, even when made of very good conductors, offers some resistance to the flow of electricity. Good conductors have low resistance and poor conductors have high *resistance*. Before defining resistance we need to look at Ohm's law.

In 1826 Georg Ohm carried out some experiments and found that if different potential differences are applied across a piece of wire then:

the current passing through the wire at a given temperature is in direct proportion to the potential difference between its ends.

This is Ohm's law, which put arithmetically says:

$$\frac{\text{potential difference}}{\text{current}} = \text{constant}$$

This constant is the resistance of the wire and is measured in ohms (Ω). A resistance of 1 ohm allows a current of 1 ampere to flow in a wire when there is a potential difference of 1 volt across it. Using the conventional symbols of I for current in amperes, V for potential difference in volts and R for resistance measured in ohms one can express Ohm's law in the following formulae:

$$\frac{V}{I} = R$$

$$I = \frac{V}{R}$$

and V = IR

Given any two of the three variables current, voltage and resistance, one can calculate the third. If, for example, one had an electric fire drawing 5 A on 240V, we can calculate from $R = \frac{V}{I}$ that the resistance is $\frac{240}{5} = 48\ \Omega$.

There are various factors, apart from the material from which a conductor is made, which influence its resistance.
a) Cross-sectional area. The smaller the cross-sectional area of a wire or cable the greater its resistance. Going back to our circuit the filament of a light bulb is very thin and its resistance high compared to the wires in the rest of the circuit.
b) Length. The longer the conductor the greater its resistance.
c) Temperature. For most conductors the hotter they get the higher their resistance becomes. Remember that heat can be generated by a current flowing in a wire, so this can in turn affect resistance.

All three factors must be taken into account when selecting a suitable conductor for a particular purpose.

Resistance can be measured using an *ohmeter*. This is often of use to the refrigeration engineer, who can use it to check the condition of a circuit – most especially, of an electric motor. Manufacturers of motors issue details of acceptable resistance values, against which one can check a suspect motor; and if there is a complete break in a motor winding – if it is open circuit – an ohmmeter will register infinite resistance to the flow of current. This proves beyond doubt that the motor has to be replaced or rewound.

Power

Electrical *power* (the rate of doing work), like other forms of power, is measured in watts. The power of a piece of electrical equipment can be calculated by multiplying the current passing through it by the potential difference across it:

$$W = IV$$

where W is the power in watts, I is the current in amperes and V is the potential difference in volts.

An electric fire on a 240V supply would, if it had a current of 5A flowing though it work at

$$240 \times 5 = 1200\ W$$

This can be expressed as 1·2 kilowatts (kW). As the power rating in watts is a rate of doing work one can easily calculate how much energy is consumed over a given time in joules. However it is more convenient, in many cases, to express energy consumed in terms of kilowatt hours (kWh). If the electric fire mentioned above were to run for one hour it would consume 1·2 kWh.

The unit of power in the British system was horsepower. (See Chapter 14 for conversion factors between watts and horsepower.)

Magnetism and electricity

As we mentioned earlier magnetism is extremely important in generating electricity. A current is generated when a wire in a circuit is moved across magnetic lines of force. If you take a magnet, put a piece of plain paper over it and then sprinkle iron filings on the paper the filings will take up a pattern and this pattern gives you a picture of the flux lines of a magnet. These are lines of magnetic force which flow from one pole of a magnet to the other in an unvarying pattern and which never cross.

The relationship between electricity and magnetism was demonstrated by Michael Faraday in 1831. He showed that if a magnet is moved through a coil of copper wire, an electric current flows through the conductor (the wire) providing it is connected to a circuit. The size of the current varies with the speed of the magnet and the number of turns of wire in the coil. Faraday could equally well have moved the coil of wire over an unmoving magnet – whichever way the experiment is conducted, a current is induced in the conductor. In more detail, an e.m.f. is generated in the wire and if this is connected to an external circuit, a current flows.

It is worth noting that the reverse also happens: an electric current can create a magnetic field around the conductor. If a current is passed through a solenoid coil, a magnetic field is induced which is identical with the natural magnetic field illustrated in Fig. 3.8. The strength of a field is considerably increased if the coil is wound around a bar of iron and this fact is made use of in the solenoid valve, described in Chapter 6. This control has the iron bar mounted vertically and connected to a spindle which has at its base a needle which opens or closes the seat of a valve regulating the flow of refrigerant, water or steam. If the solenoid is energised the induced magnetic field is strong enough to lift the spindle and needle by magnetic force; the valve is opened, and the medium which it controls can flow through the system. If the solenoid is de-energised and

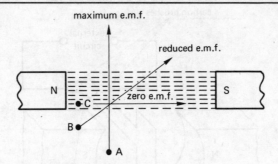

Fig. 3.9 Graph showing the e.m.f. induced in a coil turned in a magnetic field

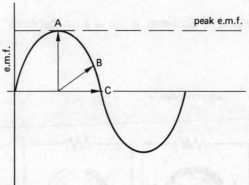

Fig. 3.10 Sine wave

the magnetic field removed, the spindle falls by gravity to seal the port and shut off the flow of liquid.

But now let's get back to the generation of currents. If Faraday's experiments are taken further, it will be seen that the strength of the induced e.m.f., which drives the current, is greatest when the coil's movement is parallel to the flux lines.

This is illustrated in Fig. 3.9. It is not necessary to move a conductor into, across and out of lines of magnetic force to create an e.m.f. If a loop of copper wire is turned around in a magnetic field, an e.m.f. is induced and its strength and direction will change as the coil rotates. The strength of the e.m.f. (and, hence, the induced current) will, if plotted, follow a sine wave (see Fig. 3.10).

Generators

And that, basically, is all that is needed to generate electricity. The loop of wire shown in Fig. 3.11 can be taken as a schematic representation of the *armature* of a *generator*. As the loop turns a current will be induced. When the coil is moving through the first half-turn the current will flow in one direction. When it starts to

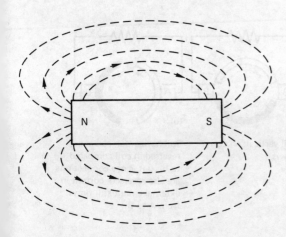

Fig. 3.8 Bar magnet flux lines

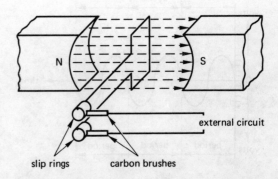

Fig. 3.11 Elementary a.c. generator – coil in a magnetic field

move through the second half-turn the direction of the current will be reversed. The current generated will be *alternating current* (a.c.).

The properties of the current shown in Fig. 3.10 can now be plotted more accurately, to indicate both its strength – on the vertical scale – and the speed with which the coil is rotated (see Fig. 3.12). That speed is expressed in terms of cycles per second for each full revolution. Hertz (Hz) is the usual abbreviation.

The current induced in the coil of an a.c. generator can be fed, through slip rings in contact with the coil, to carbon brushes connected to external conductors. If d.c. – *direct current* – is required, one of the slip rings is omitted. The second is split, and one end of the coil connected to each half of what is now called a *commutator*. The brushes are placed opposite each other, and as the armature loop is rotated, the commutator automatically switches each

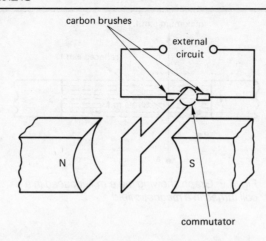

Fig. 3.13 Schematic layout of a d.c. generator

half of the loop from one brush to another every time the coil turns through 180°. So, although the current flow in the copper wire is reversed every half-turn, the current in the external circuit is not reversed and one has a direct current (d.c.). The layout of a d.c. generator is shown in Fig. 3.13 and split rings are shown in Fig. 3.14.

The armature of a generator must of course be turned at a precise speed. The prime mover is usually a turbine, driven by steam generated by oil, coal, or nuclear energy; or perhaps turned by a flow of water. *Note*: in large a.c. generators the coil is stationary and the magnet rotates inside it.

Before being fed into the distribution grid system, current is stepped up to a high voltage by a *transformer*. Transformers are

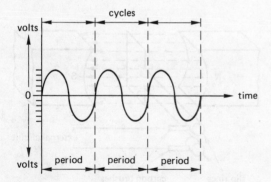

Fig. 3.12 Variation of alternating current with time

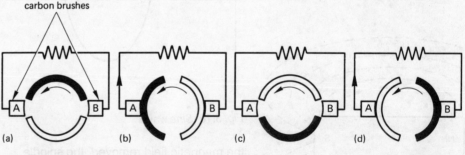

(a) Coil vertical; momentarily no current in coil and no potential difference between A and B.

(b) Coil horizontal; current flowing in coil and from A to B.

(c) Coil vertical; no current flows.

(d) Coil horizontal; current reversed in coil compared to b), but, since connections are also reversed, current in external circuit still flows from A to B.

Fig. 3.14 Split rings

ELECTRICITY – THEORY

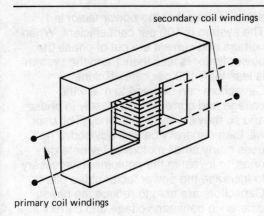

Fig. 3.15 Single-phase transformer

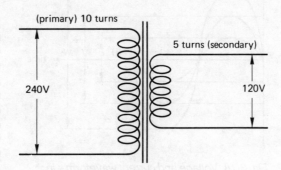

Fig. 3.16 Transformer coil ratio – turns: voltage

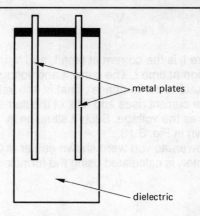

Fig. 3.17 Capacitor cross-section

used at the other end to step the voltage down to levels convenient and safe for domestic use. The construction of a transformer is shown in Fig. 3.15.

Transformers

Transformer coils are normally wound onto a laminated iron core. The conductor carrying the incoming current – the *primary coil* – is not connected to the *secondary coil*, but the two coils are linked together magnetically so that a current flowing through the primary coil induces a current in the secondary coil. What, in fact, happens is that the current in the primary coil causes an alternating magnetic flux in the iron, which in turn induces a current in the secondary coil. An a.c. primary current will induce an a.c. secondary current and the voltage of the two currents will be in ratio to the number of turns of wire each one contains.

$$\frac{\text{e.m.f. in secondary coil}}{\text{e.m.f. in primary coil}} = \frac{\text{number of turns in secondary coil}}{\text{number of turns in primary coil}}$$

If the incoming e.m.f. is 240V, and the primary coil has 240 turns of wire whilst the secondary coil has only 120, the induced e.m.f. will be 120V – half that of the primary voltage. Transformers can be used to step down or to step up voltages by changing the ratios of the coil loops.

Capacitance

Capacitors (sometimes called condensers, which can be very misleading since the same name is also given to the external heat exchanger in a refrigeration system) have the ability to store electric charge. They consist, essentially, of two metal plates separated by a dielectric as shown in Fig. 3.17. Their capacitance (their ability to store electric charge) depends upon the area of the plates, the distance apart of the plates and the nature of the dielectric.

Capacitance is the ratio of the charged stored to the potential difference across its plates and is measured in farads (F). If a capacitor holding a charge of 1 coulomb has a potential difference across its plates of 1 volt it has a capacitance of 1 farad. One farad is a very large value for capacitance and it is more usual to work in terms of microfarads (μF). One microfarad is one millionth of a farad.

We use capacitors for two purposes: to increase the starting *torque* of a single-phase a.c. motor and to maximise the efficiency in an a.c. circuit.

Maximising efficiency

When an a.c. power supply is put across a resistor a current flows in the resistor and at any time, the value of the current can be given by the formula:

$$I_t = \frac{V_t}{R}$$

where I_t is the current at time t, and V_t the voltage at time t. The current and voltage are said to be 'in phase'. That is, the value of the current rises and falls at the same time as the voltage. Such a situation is shown in Fig. 3.18.

Power, as you were shown earlier in this chapter, is calculated using the formula:

$$W = IV$$

The cost of producing electricity depends on the mains voltage and the current, so one pays for electricity on this basis. The power consumed is taken to be mains voltage multiplied by current (VA). When the voltage and current are in phase VA is equal to the power actually used (the active power).

However, things are not so simple with generators and motors. In this case the voltage and current are not in phase, but 'out of phase' as shown in Fig. 3.19. When voltage is at its peak, the current is low and when current is at its peak voltage is low. This leads to the active power (that actually used) being less than VA, but the cost to the supply authorities depends on VA rather than active power. Therefore they charge industrial companies with large motor loads on a two-part tariff. The first part covers the cost of the active power, as for any supply. The second part of the tariff is based on a maximum demand meter which is used to estimate VA. To reduce his bill to a minimum the industrial user should therefore try to keep VA to a

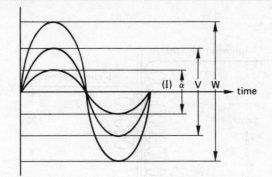

Fig. 3.18 Voltage and current waveform – in phase

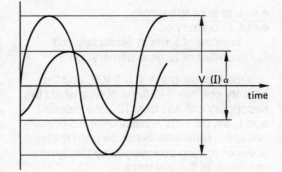

Fig. 3.19 Voltage and current waveform – out of phase

minimum by having a 'power factor' as near to unity as possible.

Power factor
When current and voltage are in phase the power used and VA are the same:

$$\frac{\text{active power}}{\text{VA}} = 1.0$$

and one says that the *power factor* is 1. The system is 100 per cent efficient. When voltage and current are out of phase the power factor is less than 1 and the system is less than 100 per cent efficient. Capacitors can be used to make the voltage and current more closely in phase and so they increase efficiency. The user will then generally be well advised, if he uses many small motors or fewer large ones, to invest in the equipment necessary to increase the power factor of the plant. Capacitors are able to reduce the phase difference between voltage and current if installed in parallel with the load to be corrected **either**
a) across individual motor terminals, **or**
b) in a bank, close to the mains transformer and switchgear.

The last word rests, of course, with the supply authority, who will often define the lowest power factor it is willing to accept.

Electrical motors

Like generators, motors also exploit the principles of electro-magnetism. However in a motor electrical energy is converted into mechanical energy, whereas in a generator mechanical energy is converted into electrical energy.

In a d.c. motor the electricity is fed into the rotating coil by a commutator and the coil will rotate between the poles of a field magnet. In fact a d.c. machine will act both as a motor or as a generator without alteration.

However a.c. motors are quite different, and although commutator motors are used for small devices, industrial motors are specialised machines in their own right.

In an a.c. motor a current is fed into a static coil – a *stator*. This sets up a magnetic field which in turn induces a current in an inner coil, which is made to rotate. This *rotor* most often has a laminated steel core, into which are inset copper bars joined to end-rings to form a 'cage' like the one shown in Fig. 3.20.

Three-phase a.c. motors

Most industrial premises are wired for three-phase supply, that is, there are three cables supplying electricity. In this case the stator will be arranged in three windings, which will be spaced at 120° intervals. Figure 3.21 illustrates how this would be shown on an electrical diagram. The shape of the coil arrangement – a Greek 'Delta' – gives it its name. When a current flows through the windings, it will induce a magnetic field, which will rotate around the triangle as each phase in succession reaches peak strength and then reverses its polarity. This will induce currents in the conductors of the rotor, which in turn produce their own magnetic fields. The interaction of these magnetic fields causes the rotor to 'chase' the rotating field until the motor reaches its maximum speed. This, in terms of revolutions per minute, can be calculated from the formula:

$$\text{r.p.m.} = \frac{\text{Cycles/sec} \times 60}{\text{No. of pairs of poles}}$$

Thus on 50 Hz supply, a two-pole motor has a maximum speed of 3 000 r.p.m., and

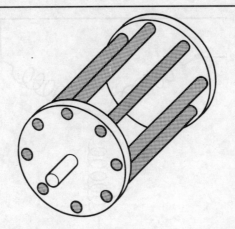

Fig. 3.20 Squirrel cage rotor

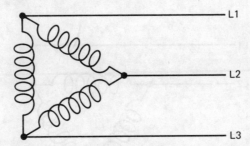

Fig. 3.21 Delta motor connections

a four-pole motor is capable of half that speed. This maximum theoretical r.p.m. is termed *synchronous speed*. In practice, it cannot be maintained; and motor manufacturers settle for between $1\frac{1}{2}$ per cent to 4 per cent less.

Squirrel cage motors

The cage construction illustrated in

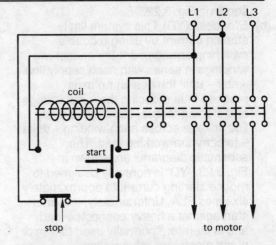

Fig. 3.22 Direct-on-line starter

Fig. 3.20 is economical to make, robust, and has a good power factor on full load – something like 0·9. On the other hand, it has a low starting torque and a high starting current – something approaching 10 times full load amps (FLA). Starting methods include provision to limit the starting current inrush in order to prevent the operation of other electrical equipment powered from the same supply source being affected.

Common starting methods include the following.

a) *Direct-on-line* In this method mains current is connected direct to the motor winding terminals. Nothing is done to reduce starting amperage, and the method is not normally acceptable to supply authorities in sizes larger than 5·6 kW ($7\frac{1}{2}$ hp). The schematic arrangement of motor and starter is shown in Fig. 3.22.

b) *Star-delta* (YD) This system limits starting current by using a double switching arrangement to connect two windings in series with each supply line – star – until the design r.p.m. is reached thus reducing the initial voltage and starting torque, and then to apply line voltage across each winding – delta – to carry the working load. The schematic diagrams are shown in Fig. 3.23. 'YD' is normally designed to reduce starting current to approximately six times FLA. Unfortunately, it cannot start against a heavy connected load, and is therefore normally used for fan or pump motors, or refrigeration compressors with provision for unloaded starting.

c) *Auto-transformer* Here, an automatically disconnected transformer is connected to a Delta winding arrangement to reduce starting current by increasing the voltage applied in stages of, for example, 45 per cent to 60 per cent to 75 per cent of line voltage, and then drop out of the circuit. The schematic diagram is shown in Fig. 3.24. This is an extremely efficient, but unfortunately expensive arrangement. For that reason it is used mainly with large capacity motors – including those in hermetically sealed compressor assemblies, especially centrifugal motors of 115 kW (150 hp) and above.

d) *Part wind* This is the most widely used starting method for American accessible hermetic compressors in the 7·5 to 115 kW (10 to 150 hp) size range. The

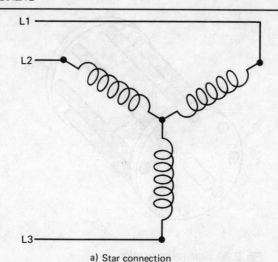
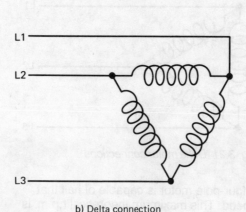

Fig. 3.23 *Star and delta motor connections*

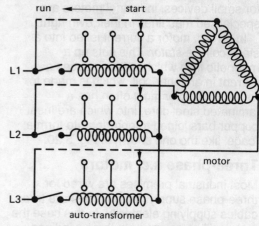

Fig. 3.24 *Schematic auto-transformer starter and motor*

motor is wound with two separate coils. The first (start) winding brings the motor up to operating speed; and the second (which is brought in magnetically between 1 and 3 minutes later) then takes the balance of the full load. Unloaded, or part-unloaded, starting characteristics are desirable, as too heavy a connected load will prevent the motor from starting until the second winding is energised, thus frustrating attempts to minimise the starting current inrush. This type of starter is therefore closely matched with its motor and compressor characteristics.

e) *Slip ring* This system cannot be operated with a cage-type rotor, and requires a wound rotor. The current admitted to the rotor windings is

ELECTRICITY – THEORY

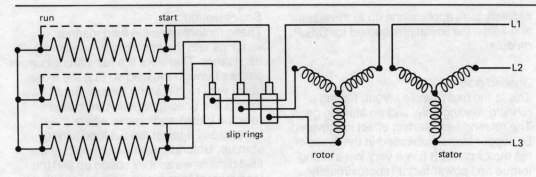

Fig. 3.25 Schematic slip-ring starter and motor

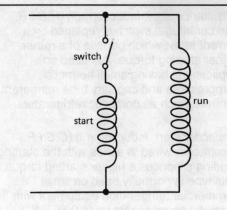

Fig. 3.26 I.S.R. motor

governed by slip rings. It is illustrated schematically in Fig. 3.25. The system is very effective in that it not only enables motors to be started under a heavy load, but also controls starting speed. Its disadvantage is the use of brushes with the slip rings, since they cannot be used in *hermetically* sealed compressor assemblies. For this reason, refrigeration engineers normally meet this type of equipment only on large open-compressored installations, including those using ammonia.

Single-phase a.c. motors

Three-phase motors are simpler and have higher starting torque, than single-phase equivalents, the latter being used principally because of the high cost of running three-phase supplies to non-industrial premises.

Starting single-phase motors is obviously more difficult, since there are no inherent rotating magnetic fields to be 'chased' by the rotor. In the great majority of cases – certainly all applications with any degree of starting torque to be provided – two windings (start and run) are necessary. The run winding remains in circuit all the time and the start winding is cut out once the motor picks up speed. The current in the start winding is approx 90° out of phase with that in the run winding and the two magnetic fields never have the same polarity. This feature gives the rotor something to 'chase' when the motor is being started; and was the origin of the term *split phase*. Single-phase types of interest to the refrigeration specialist include the following, illustrated in Figs 3.26–3.31.

Induction start, induction run (I.S.R.)
The starting winding is in series with a centrifugal switch, which breaks circuit when the motor reaches its design running speed. This type has only low starting torque and is used with fans or other equipment which start virtually unloaded.

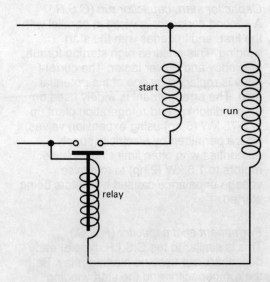

Fig. 3.27 R.S.I.R. motor

27

Resistance start, induction run (R.S.I.R.)
The centrifugal switch is replaced by a *current relay*, which permits of a rather higher starting torque. It is used on applications having small hermetic compressors and capillary tube refrigerant controls, such as domestic refrigerators.

Capacitor start, induction run (C.S.I.R.)
A capacitor wired in series with the starting winding produces a higher starting torque. This type is normally used on small commercial refrigeration equipment with hermetic compressors up to 0·55 kW (¾ hp).

Capacitor start, capacitor run (C.S.R.)
A second capacitor is wired in parallel with the first, and in series with the start winding. This ensures high starting torque, efficiency and power factor. The current relay is replaced by one of the potential type. The arrangement is widely used on air-conditioning and refrigeration plant up to 3·75 kW (5 hp) using expansion valves, where permitted by electricity supply authorities, who often limit single-phase motors to 1·5 kW (2 hp) to minimise voltage unbalance caused by motors being started.

Permanent split capacitor (P.S.C.)
This is similar to the C.S.I.R. type already described, but uses no starting relay. Both the run capacitor and the start winding remain in circuit after the motor reaches running speed. This system was developed for air-conditioning applications, and is used on almost all 'window' units and other capillary tube applications up to the sizes and within the limitations quoted for C.S.R. models.

Shaded pole
This is the most simple layout, having a running winding only, and no starting gear. The rotating field starting effect is provided by copper loops embedded in the faces of the motor poles. It has a very low starting torque and power factor (approximately 0·6) at full load, and low efficiency. It is normally used only for fan motors, in this industry.

D.C. motors
D.C. equipment is not widely used in other than marine and specialist applications, such as lift and crane motors. The need to use brushes prevents d.c. motors being used in hermetic compressors, and they are infrequently seen by the average refrigeration fitter. A pity, since d.c. motors do not have any phase relationship complications to worry about! There are three main types in widespread use.

Shunt wound
In this type both the armature and the field magnet windings are connected in parallel (shunted across) the supply. There is little loss of speed – about 5 per cent – when load is applied. The starter reduces armature current by means of a variable resistance in the armature circuit, until full speed is reached and is an example of a 'face plate' motor starter. The schematic diagram is shown in Fig. 3.32.

Series wound
These motors have the field magnet windings and the armature in series with each other. Therefore the full starting current passes through the field so that the motor has a very high starting torque. Such motors are used for traction purposes (electric trains) and cranes where they have to pull against the full load as they start up. Unfortunately the strength of the field diminishes as they speed up and the load current is reduced. This causes them to speed up even further and, if the load is accidentally removed, they over-speed which is dangerous. The starting current is restricted by a simple, variable resistance that is also used as speed controller. Unlike the shunt wound motor which is a constant speed device, series wound motors are used for variable speed applications such as electric vehicles.

Compound wound
These motors, shown in Fig. 3.34, are the type most often used in refrigeration work, and contain both series and shunt fields. This provides a higher starting torque than for a simple shunt wound motor and by balancing the two windings various load/speed characteristics can be produced. A face-plate starter is used to limit the starting current to a safe level. It is most often used with open-type compressors used aboard ship.

Variable resistance starters
These starters normally incorporate two safety devices: overload and no-volt trips, and the layout of a typical starter for use

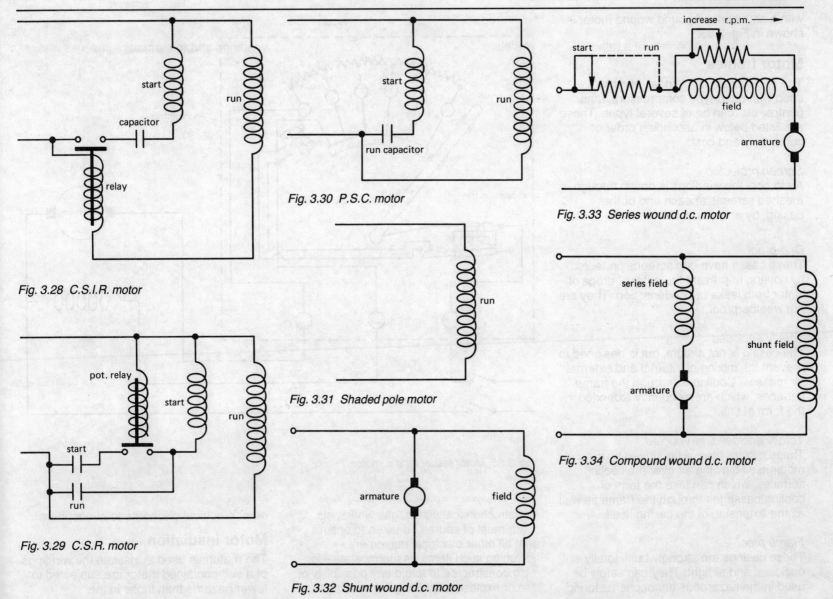

Fig. 3.28 C.S.I.R. motor

Fig. 3.29 C.S.R. motor

Fig. 3.30 P.S.C. motor

Fig. 3.31 Shaded pole motor

Fig. 3.32 Shunt wound d.c. motor

Fig. 3.33 Series wound d.c. motor

Fig. 3.34 Compound wound d.c. motor

TROPICAL REFRIGERATION AND AIR-CONDITIONING

with a shunt or compound wound motor is shown in Fig. 3.35.

Motor frames

The enclosures, or 'frames', of motors used with open-type compressors, fans, pumps, etc. can be of several types. These are listed below in ascending order of complexity and cost.

Screen protected
Air to cool the windings is drawn through meshed screens at each end of the casing, by an internal fan.

Drip-proof
These cases have end screens protected by covers, to prevent the entry of drops of water from leaks or condensation. They are *not* weatherproof.

Totally enclosed
The casing is not airtight, but is designed to prevent the mixing of internal and external air masses. Cooling is through the frame surfaces, which are frequently extended in the form of fins.

Totally enclosed, fan-cooled
These motors have a fan driven by the motor itself blowing air over the cooling surfaces, which can take the form of cooling passages through the frame as well as the extension of the casing itself.

Flame proof
These casings are strongly built, totally enclosed and airtight. They can safely be used in the hazardous atmospheres found in petrochemical plants, gas works, etc. Care must of course be taken to ensure that all other electrical equipment – including such items as thermostats – is also constructed to avoid any possibility of fire or explosion resulting from arcing or other electrical discharge to atmosphere.

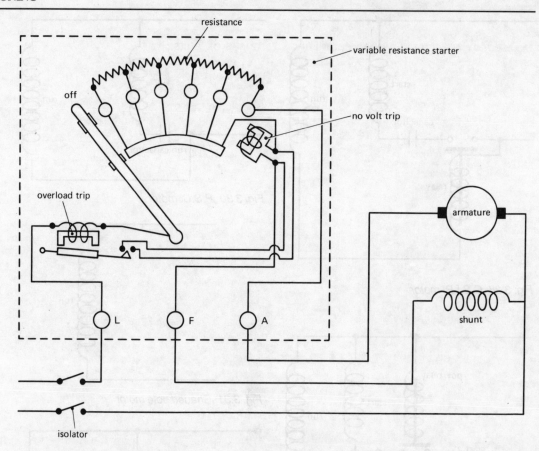

Fig. 3.35 Motor starter for d.c. motor

Motor insulation

The materials used to insulate the windings of a self-contained motor are subjected to fewer hazards than those in the

30

ELECTRICITY – THEORY

stator–rotor units of hermetically sealed compressors, which are constantly exposed to the action of the refrigerant and its entrained oil. In the majority of cases the refrigeration man will find stator–rotor unit windings insulated with specially developed varnishes, which are impervious to the halogen refrigerants in their gaseous and liquid forms.

Insulation ratings are indications of the extent of temperature increase which can be accepted without damaging the insulation. Fig. 3.36 shows the common standards.

Fig. 3.36 Motor insulation temperature limits

Class	°C
A	105
E	120
B	130
F	155
H	180
C	180+

N.B. The above limits include a coolant temperature of 40 °C which must be subtracted from the limiting temperature to give the maximum temperature rise permissable. For example, class B motor rise is: 130 − 40 = 90 °C.

Detailed instructions for testing windings for both insulation and for continuity are contained in the service instructions in Chapter 10 – 'Trouble Shooting' – and service requirements generally are covered under notes on specific products. However the general notes on a.c. and d.c. motor and starter problems in Fig. 3.37 provide an outline of possible symptoms and causes of faults.

Power distribution

The sketches used in this chapter have shown only two or three conductors, but the distribution in customers' premises can take other forms described below and illustrated in Figs. 3.38–3.40:

a) Two-wire, single-phase a.c., or d.c. – domestic premises or small users.
b) Three-wire, three-phase a.c., or d.c. – larger users.

Fig. 3.37 Basic motor and starter faults

Symptom	A.C. motor	D.C. motor
Motor will not start	a) Thermal or current overload open circuit b) Faulty starting relay, capacitor or centrifugal switch c) Bad connections on controls or control gear	a) Starter resistance open circuit b) Open circuit in field winding c) Bad connections on controls or control gear
Fuses blow on starting	a) Motor overloaded or seized b) Starter defect c) Earth or short in windings d) Single phasing (one phase or line open circuit)	a) Motor overloaded or seized b) Starter defect c) Short circuit in field or armature circuit
Motor hums, will not start	a) Low supply voltage at motor terminals b) Defective capacitor (1 ph) c) Mechanical overload or seizure	a) Low supply voltage b) Mechanical overload or seizure
Motor overheats	a) Mechanical overload b) Winding defect c) Worn bearings d) Fluff, dirt or grease in motor frame	a) Mechanical overload b) Field or armature coil defect c) Worn bearings d) Fluff, dirt or grease in motor frame
Sparking at brushes	N/A	a) Incorrect brush position b) Defective armature coil c) Dirty or worn commutator d) Mechanical overload or bearing defects

TROPICAL REFRIGERATION AND AIR-CONDITIONING

c) Four-wire, three-phase a.c. – commercial/industrial users.

For our purposes, it can be assumed that power will arrive at the customer's mains distribution board at the voltage(s) of the equipment on which we shall be working. The incoming supply will be protected by mains fuses, which are the property and responsibility of the electricity supply authority, and *must not* be touched by unauthorised personnel.

Two-wire installations – a.c. or d.c.

Incoming, fused supplies will be metered, connected to a mains isolator, and then distributed to other fused isolators for each main group of functions – for example, lighting, ring mains power, and electric cookers or other appliances drawing large currents. A typical two-wire a.c. diagram is shown in Fig. 3.38. Note that from each sub-distribution board, the wiring layout to lights etc. would be the same for d.c. as well as a.c. supplies. In both cases, one conductor is positive, and the other a neutral (sometimes called 'negative') line.

Three-wire d.c.

When this arrangement is used, two conductors are positive and the third a neutral line. This enables lower (line) voltages and higher voltages to be supplied within one circuit (Fig. 3.39). Note that the neutral line is effectively earthed.

Four-wire, three-phase a.c.

In this arrangement, the fourth conductor is used to provide a neutral line. Each phase

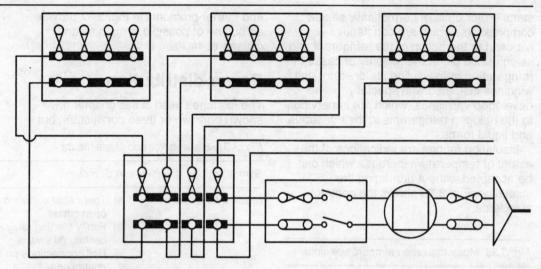

Fig. 3.38 2 wire, single-phase a.c. circuit

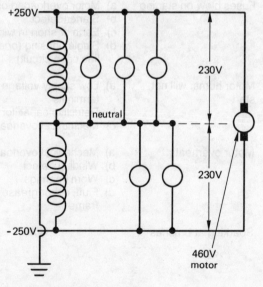

Fig. 3.39 3 wire d.c. circuit

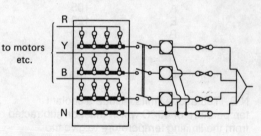

Fig. 3.40 4 wire, 3 phase a.c. circuit

ELECTRICITY – THEORY

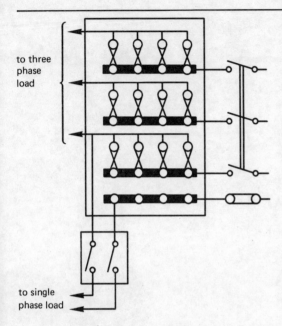

Fig. 3.41 Single-phase tap-off

is fused and metered before being wired to a T.P. & N. (Triple Pole and Neutral) fused isolator, and distributed as required (see Fig. 3.40).

As Fig. 3.41 illustrates, any single-phase requirements can be met by current from one of the three-phase conductors, using the neutral line to complete the circuit. Each single-phase circuit so arranged must be taken through a separate fused isolator and where several such loads are supplied consumption should be balanced as closely as possible between all three phases.

Earthing

Although *earthing* is discussed only briefly, it is a major safety factor.

The word means what it says – providing a means of conducting any 'stray' electricity, escaping from a damaged cable for example, directly and safely to the earth. The most effective earth is provided by a deeply buried copper rod or pipe, to which earthing connections are made from the neutral lines of single- or three-phase circuits. The practice of earthing to a water pipe is *not* recommended, since it cannot be guaranteed to safeguard anyone touching a damaged piece of equipment.

Earth-leakage circuit breakers are used in many large installations. If a live lead touches the earthing conductor, the circuit breaker (C.B.) operates to trip the incoming mains supply. The C.B. can be either of the current-or voltage-operated type, and will have a manual reset button. If this has tripped, a piece of equipment or a cable within the installation will present a potentially lethal hazard if the supply is reconnected without finding and repairing the defect. **This must be done only by properly qualified electricians**.

Colour coding of conductors

The identification of conductors is helped by the use of colour coding of the conductor insulation. In the E.E.C., the following are required standards.
a) Flexible cords
positive = brown
neutral = blue
earth = yellow and green
b) Power cables a.c. 1-phase or d.c.
positive = red
neutral = black
                    a.c. 3-phase lines
supply = red
supply = yellow
supply = blue
neutral = black.

Manufacturers' wiring diagrams should indicate the colours of all conductors, and the identity or function of all control equipment and connections. Similar information should be included in 'as fitted' drawings covering on-site installation work.

The Peltier effect

It is appropriate to end our notes on electrical theory with a brief description of an electronic refrigeration circuit. This ties up the loose ends of an earlier observation that a current can be generated by heating joints between two dissimilar conductors. The reverse is also true – if an electric current is passed through such a circuit, heat will be evolved at one side of the couple, whilst the other side will become cooler. This is illustrated schematically in Fig. 3.42, and is known as the Peltier effect.

With the development of semiconducting materials and transistors, this effect has been used to cool small refrigerators used on such applications as hotel bedroom. drink cabinets. So far, however, the materials used have not proved sufficiently efficient and long lived to make electronic

TROPICAL REFRIGERATION AND AIR-CONDITIONING

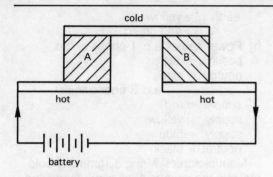

Fig. 3.42 Peltier effect

refrigeration a commercially attractive proposition. It may be that suitable materials will have to be developed, rather than found by trial and error, in the same way that modern refrigerants were developed by Thomas Midgely. Until they are available, the Peltier Effect seems doomed to remain an interesting theory rather than a practical refrigeration method.

Revision questions

1. a) State the present British Standard colour code for each of the following in a single-phase wiring system:
    (i) phase,
    (ii) neutral,
    (iii) earth.
   b) A domestic refrigerator contains a hermetically sealed compressor, a magnetic relay, an overload protector and a cabinet bulb.
   Sketch the wiring diagram based on the British Standard colour code for single-phase wiring.
   c) A new refrigerator has just been purchased from America and the voltage rating is 110 V. What instrument could be recommended for the refrigerator to work efficiently in Nigeria?
   (WAEC)

2. Name three measuring instruments commonly used on electrical circuits. What does each measure?

4. Refrigerants

The majority of domestic and commercial refrigerators and air-conditioners in current use are charged with one of two refrigerants: R12 for refrigeration, and R22 for air-conditioning applications. Examination of larger or more specialised installations will however reveal several others in common use. They include R717, ammonia, in block ice-making plants, or factory-scale food freezing and cold-storage equipment. In large centrifugal compressors operating at relatively low speeds – 3 000 r.p.m., say – R11 is often used. But for really high speed centrifugals–running at perhaps 23 000 r.p.m. – we would probably find R12 being used.

New refrigerants are developed to get the best possible results from new types of equipment, or to boost the performance of existing compressor designs – perhaps because of a need to use them at higher or lower evaporating pressures. Several examples of specialised usages are quoted in the following notes. These include information on refrigerants not yet widely used outside the USA and Europe, but likely to be more commonly used in the foreseeable future.

In addition to names and chemical formulae, relative pressures and relative efficiencies are noted, and we include a brief summary of the types of application on which each refrigerant is most likely to be encountered. We also note any special characteristics which may be of general interest, or enable you to handle them more safely.

Safe working procedures must not only be followed, but be understood by everyone working with refrigerants.

Storage and safety

Most refrigerants are supplied, and stored, in large pressure vessels holding, say, 60 kg (132 lbs) of liquid and vapour when full. These must be stored upright, with caps in place, in cool and well ventilated stores located well away from boiler rooms, or areas in which operations presenting fire hazards – welding, for example – are carried out. Similarly, when cylinders are used in the workshop or on site, brazing torches or welding sets *must not* be used close to them. Given a pressure cylinder containing liquid and vapour, plus a source of intense heat and just a little bad luck, one has all the ingredients necessary to cause a lethal explosion.

The requirement for storage of cylinders in well ventilated rooms covers both the need to remove any refrigerant which might escape, and the need to keep the cylinders as cool as possible. In practical terms, storage area temperatures should not reach levels at which excessive refrigerant pressure is generated; and an upper limit of not more than 2 070 k Pa (300 psi) is recommended.

At all times bear in mind that refrigerants have been specially developed to remove a lot of heat quickly through any surface on which they boil. If liquid refrigerant sprays onto your skin, you'll learn all about 'cold burns', or frostbite. The chapter on first aid includes advice on the immediate treatment of cold burns, as well as hot burns. Reading it, and remembering what to do in the event of an accident may prove a good investment, but prevention is better than cure! Never take chances with refrigerants – if you have to open a circuit, or purge it, make sure that liquid cannot spray onto your hands and arms or, even worse, into your face and eyes. If possible, pipe it away to the outside air (except in the case of R717, which should be released under the surface of a large drum or tank of water). Don't think it cowardly to wear safety glasses and gloves. And never fool around with *any* liquid or gas stored under pressure; even air can kill if its pressure enables it to find its way into the bloodstream.

Refrigerant cylinder valves

All cylinders are fitted with back-seating type valves screwed directly into the

cylinder necks, and protected by heavy caps. These valves are either open, or closed:
a) when the valve is front seated it is closed and the refrigerant is sealed within the cylinder
b) when the valve is back seated, the charging port will be open, and refrigerant will flow through it.

Standard valve outlets are ¼″ SAE fittings which should be kept sealed and capped when the cylinder is not in use with the valve cap in place to avoid accidental damage.

'Canned' gas valves

The most popular fluorocarbon refrigerants are also available in small cans holding about ½ or 1 kg of liquid. These do not have valves and require the use of special tapping valves which incorporate base locking adaptors. These should be securely clamped to the top of the can and the valve body screwed in clockwise until the can is pierced by the tapping device. Like large cylinders these can valves have ¼″ SAE connections for charging hoses; and most – there is a wide range of designs and costs – have a manually operated shut-off valve. The cans are disposable, but do not throw away the valve with them! Properly looked after, the valves can have a long life.

Cylinder colour codes

Whilst the refrigerant number code is clear and simple, the chemical formulae and names of refrigerants are not. A colour

a) Universal type   b) Spud-can type

Fig. 4.1 'Canned' gas valves

code for the quick identification of the contents of cylinders was therefore introduced by du Pont. Learn it, and check it out each time you use a fresh cylinder of refrigerant.

Refrigerant properties

An ideal refrigerant would meet many requirements. It should not be toxic or flammable, and would mix with oil. It would have a low boiling point, and a condensing temperature low enough to require a minimum degree of compression. A small volume of the perfect refrigerant would, by virtue of its high latent heat content, be capable of a high cooling duty. The compressor displacement for a given refrigeration requirement would be kept to a minimum. It would function at relatively low pressures. It would be cheap. And unfortunately it could not exist in one single form, since so many ranges of temperature have to be catered for by both evaporators and compressors!

The refrigerants which you are most likely to use, or to encounter, are listed below with their basic data. First an explanation of some of the terms used is necessary.

In the list and in other places in the book you will find the abbreviation 'TR'. This stands for 'ton of refrigerating effect' and means the refrigerating effect of 1 ton of ice melting over a period of 24 hours; 1 TR is equal to 3·5 kJ/s (12 000 Btu/h).

The abbreviation 'cfm per 1 TR' stands for 'cubic feet per minute per ton of refrigerating effect' (at set conditions). It therefore shows the rate of flow of refrigerant needed for a given cooling duty.

The coefficient of performance is defined as: the refrigerating effect (energy absorbed at the evaporator) divided by work (energy required by the compressor) with both energy figures in the same units.

The UL Hazard Class refers to the standards laid down in America by the Underwriters' Laboratory. The classes give a comparative indication of safety. The classifications starts at 1, for the most hazardous chemicals, and works its way towards the safest refrigerants available through categories 2, 3, 4, 4–5, 5, 5a, 5b and – the highest standard – 6. Whilst the refrigerants listed are all very safe in relative terms, breathing them in high concentrations is dangerous: your lungs need oxygen! Also, if they are exposed to fierce heat which breaks down their chemical structure, acids and poisonous fumes may be formed. That is not a usual occurrence, and fluorocarbons can be safely exposed to steady temperatures ranging from 107 °C (225 °F) to 149 °C (300 °F).

R11 Formula: CCl_3F Chemical name: Trichlorofluoromethane
Boiling point: 23·8 °C (74·9 °F)
At −15 °C (5 °F) suction and 30 °C (86 °F) condensing temperatures:
condensing pressure = 3·5 psig
hp per 1 TR = 0·94
cfm per 1 TR = 36·54
coefficient of performance = 5·03
Colour code: orange
UL Hazard Class: 5a
Notes: Has a faint odour; non-flammable; a low pressure refrigerant used mainly in large low speed centrifugals; an excellent solvent, much used to wash out systems contaminated by motor burnouts; supplied in low pressure drums, or cylinders pressurised with nitrogen as a propellant.

R12 Formula: CCl_2F_2 Chemical name: Dichlorodifluoromethane
Boiling point: −30 °C (−21·6 °F)
At −15 °C (5 °F) suction and 30 °C (86 °F) condensing temperatures:
condensing pressure = 93·3 psig
hp per 1 TR = 1·00
cfm per 1 TR = 5·83
coefficient of performance = 4·70
Colour code: white
UL Hazard Class: 6
Notes: Faint odour; non-flammable; the most widely used of all refrigerants, notably for domestic and commercial refrigeration applications, and automobile air-conditioning; available in various sizes of pressure cylinder and also small throw-away cans used to field charge small systems.

R22 Formula: $CHClF_2$ Chemical name: Chlorodifluoromethane
Boiling point: −40·8 °C (−41·4 °F)
At −15 °C (5 °F) suction and 30 °C (86 °F) condensing temperatures:
condensing pressure = 158·2 psig
hp per 1 TR = 1·01
cfm per 1 TR = 3·55
coefficient of performance = 4·66
Colour code: green
UL Hazard Class: 5a
Notes: Has a faint odour; non-flammable; the most used refrigerant for air-conditioning applications with reciprocating compressors; also used in some medium and low temperature refrigeration equipment to secure greater performance than R12 from a given compressor displacement.

R114 Formula: $CClF_2CClF_2$ Chemical name: Dichlorotetrafluoroethane
Boiling point: 3·8 °C (38·8 °F)
At −15 °C (5 °F) suction and 30 °C (86 °F) condensing temperatures:
condensing pressure = 22·0 psig
hp per 1 TR = 1·05
cfm per 1 TR = 20·14
coefficient of performance = 4·49
Colour code: dark blue
UL Hazard Class: 6
Notes: Faint odour; non-flammable; developed in 1933 for use in Frigidaire rotary compressors; still used in some rotary models for domestic application, and a few industrial processes.

R500 This is a mixture of R12 and R152a, technically known as an *azeotrope* or constant-boiling mixture. The formula and chemical name of R12 is as above and of R152a is CH_3CHF_2, named difluoroethane.
Boiling point: −33·5 °C (−28·3 °F)
At −15 °C (5 °F) suction and 30 °C (86 °F) condensing temperatures:
condensing pressure = 112·9 psig
hp per 1 TR = 1·01
cfm per 1 TR = 4·95
coefficient of performance = 4·65
Colour code: yellow
UL Hazard Class: 5a
Notes: previously known as Carrene 500; has a refrigerating effect some 18 per cent greater than R12, which neatly offsets the performance lost when 60 Hz

equipment is used on 50 Hz mains supply; the usual fluorocarbon-characteristics.

R502 This is an azeotropic mixture of R22 and R115. The formula of R115 is $CClF_2CF_3$, and its name chloropentafluoroethane.
Boiling point: $-45·5$ °C ($-49·8$ °F)
At -15 °C (5 °F) suction and 30 °C (86 °F) condensing temperatures:
condensing pressure = 175·1 psig
hp per 1 TR = 1·08
cfm per 1 TR = 3·61
coefficient of performance = 4·37
Colour code: orchid
UL Hazard Class: 5a
Notes: higher performance than R22, and particularly useful for reciprocating low temperature applications; as the figures suggest, condensing pressures can be very high and use in ultra-high ambient temperatures is not always possible.

R717 Formula: NH_3 Chemical name: ammonia
Boiling point: $-33·3$ °C ($-28·0$ °F)
At -15 °C (5 °F) suction and 30 °C (86 °F) condensing temperatures:
condensing pressure = 154·5 psig
hp per 1 TR = 0·99
cfm per 1 TR = 3·44
coefficient of performance = 4·76
No colour code
UL Hazard Class: 2
Notes: although an extremely efficient refrigerant, R717 is both toxic and explosive; since it attacks copper it is only used in open-type reciprocating or screw-type compressors and in systems fabricated from steel; gas masks must be available for the use of plant operators; the exception is when R717 is used with hydrogen and water, in domestic refrigerators of the absorption type, which have no moving parts, and use a hermetically sealed, heavy gauge steel circuit.

Pressure/temperature relationships

The full physical properties of refrigerants can only be detailed in extensive and complicated tables and charts, which are not easy to understand. Exhaustive details are not however important to the installation or service engineer, whose needs are concentrated on converting gauge pressures into temperatures, or temperatures into gauge pressures, for each of the refrigerants with which he works.

Simple temperature to pressure relationships are detailed in Fig. 4.2 which is straightforward in its practical application. If the installation man is balancing an R12 system to run at a saturated temperature of -29 °C (-20 °F), he aims at a gauge reading of 0·6 psig. If his discharge gauge reading is 136·4 psig, the table will tell him that the condensing temperature is 43·3 °C (110 °F). (When referring to gauge pressures the abbreviation psig is used for pounds per square inch, for absolute pressures the abbreviation is psia.)

These figures give fresh meaning to statements made earlier. You can see that R11 is indeed 'a low pressure refrigerant', and that it would not be difficult to use as a cleansing agent. We said that 'condensing pressures can be very high and use in ultra-high ambient temperatures is not always possible in our notes on R502; the table enables you to put flesh onto the bones of that statement. It indicates where, in terms of pressure, things are comfortable; and where, in terms of condensing temperatures, you need to be careful.

Fig. 4.2 Refrigerant temperature/pressure relationships

Temp °F	Gauge Pressure – PSIG							Temp °C
	R11	R12	R22	R114	R500	R502	R717	
−40	28.4*	11.0*	0.5	26.0*	7.6*	4.1	8.7*	−40
−38	28.3*	10.0*	1.3	25.8*	6.4*	5.1	7.4*	−38.9
−36	28.2*	8.9*	2.2	25.5*	5.2*	6.0	6.1*	−37.8
−34	28.1*	7.8*	3.0	25.2*	3.9*	7.0	4.7*	−36.7
−32	27.9*	6.7*	3.9	25.0*	2.6*	8.1	3.2*	−35.6
−30	27.8*	5.5*	4.9	24.6*	1.2*	9.2	1.6*	−34.4
−28	27.7*	4.3*	5.9	24.3*	0.1	10.3	0.0	−33.3
−26	27.5*	3.0*	6.9	24.0*	0.9	11.5	0.8	−32.2
−24	27.4*	1.6*	7.9	23.6*	1.6	12.7	1.7	−31.1
−22	27.2*	0.3*	9.0	23.2*	2.4	14.0	2.6	−30.0
−20	27.0*	0.6	10.1	22.9*	3.2	15.3	3.6	−28.9
−18	26.8*	1.3	11.3	22.4*	4.1	16.7	4.6	−27.8
−16	26.6*	2.1	12.5	22.0*	5.0	18.1	5.6	−26.7
−14	26.4*	2.8	13.8	21.6*	5.9	19.5	6.7	−25.6
−12	26.2*	3.7	15.1	21.1*	6.8	21.0	7.9	−24.4
−10	26.0*	4.5	16.5	20.6*	7.8	22.6	9.0	−23.3
−8	25.8*	5.4	17.9	20.1*	8.8	24.2	10.3	−22.2
−6	25.5*	6.3	19.3	19.6*	9.9	25.8	11.6	−21.1
−4	25.3*	7.2	20.8	19.0*	11.0	27.5	12.9	−20.0
−2	25.0*	8.2	22.4	18.4*	12.1	29.3	14.3	−18.9
0	24.7*	9.2	24.0	17.8*	13.3	31.1	15.7	−17.8
2	24.4*	10.2	25.6	17.2*	14.5	32.9	17.2	−16.7
4	24.1*	11.2	27.3	16.5*	15.7	34.8	18.8	−15.6
6	23.8*	12.3	29.1	15.8*	17.0	36.9	20.4	−14.4
8	23.4*	13.5	30.9	15.1*	18.4	38.9	22.1	−13.3
10	23.1*	14.6	32.8	14.4*	19.7	41.0	23.8	−12.2
12	22.7*	15.8	34.7	13.6*	21.2	43.2	25.6	−11.1
14	22.3*	17.1	36.7	12.8*	22.6	45.4	27.5	−10.0
16	21.9*	18.4	38.7	12.0*	24.1	47.7	29.4	−8.9
18	21.5*	19.7	40.9	11.1*	25.7	50.0	31.4	−7.8
20	21.1*	21.0	43.0	10.2*	27.3	52.5	33.5	−6.7
22	20.6*	22.4	45.3	9.3*	28.9	54.9	35.7	−5.6
24	20.1*	23.9	47.6	8.3*	30.6	57.5	37.9	−4.4
26	19.7*	25.4	49.9	7.3*	32.4	60.1	40.2	−3.3
28	19.1*	26.9	52.4	6.3*	34.3	62.8	42.6	−2.2
30	18.6*	28.5	54.9	5.2*	36.0	65.6	45.0	−1.1
32	18.1*	30.1	57.5	4.1*	37.9	68.4	47.6	0
34	17.5*	31.7	60.1	2.9*	39.9	71.3	50.2	1.1
36	16.9*	33.4	62.8	1.7*	41.9	74.3	52.9	2.2
38	16.3*	35.2	65.6	0.6*	43.9	77.4	55.7	3.3
40	15.6*	37.0	68.5	0.4	46.1	80.5	58.6	4.4
42	15.0*	38.8	71.5	1.0	48.2	83.8	61.6	5.6
44	14.3*	40.7	74.5	1.7	50.5	87.0	64.7	6.7
46	13.6*	42.7	77.6	2.4	52.8	90.4	67.9	7.8
48	12.8*	44.7	80.8	3.1	55.1	93.9	71.1	8.9
50	12.0*	46.7	84.0	3.8	57.6	97.4	74.5	10.0
52	11.2*	48.8	87.4	4.6	60.0	101.1	78.0	11.1
54	10.4*	51.0	90.8	5.4	62.6	104.8	81.5	12.2
56	9.6*	53.2	94.3	6.2	65.2	108.6	85.2	13.3
58	8.7*	55.4	97.9	7.0	67.9	112.4	89.0	14.4

* ins Hg Vacuum (Gauge pressures are expressed in inches of mercury if they correspond to absolute pressures of less than 1 standard atmosphere.)

TROPICAL REFRIGERATION AND AIR-CONDITIONING

Temp °F	Gauge Pressure – PSIG								Temp °C
	R11	R12	R22	R114	R500	R502	R717		
60	7.8*	57.7	101.6	7.9	70.6	116.4	92.9		15.6
62	6.8*	60.1	105.4	8.8	73.5	120.5	96.9		16.7
64	5.9*	62.5	109.3	9.7	76.3	124.6	101.0		17.8
66	4.9*	65.0	113.2	10.6	79.3	128.9	105.3		18.9
68	3.8*	67.6	117.3	11.6	82.3	133.2	109.6		20.0
70	2.8*	70.2	121.4	12.6	85.4	137.6	114.1		21.1
72	1.6*	72.9	125.7	13.6	88.6	142.2	118.7		22.2
74	0.5*	75.6	130.0	14.6	91.8	146.8	123.4		23.3
76	0.3	78.4	134.5	15.7	95.1	151.5	128.3		24.4
78	0.9	81.3	139.0	16.8	98.5	156.3	133.2		25.6
80	1.5	84.2	143.6	18.0	102.0	161.2	138.3		26.7
82	2.2	87.2	148.4	19.1	105.6	166.2	143.6		27.8
84	2.8	90.2	153.2	20.3	109.2	171.4	149.0		28.9
86	3.5	93.3	158.2	21.6	112.9	176.6	154.5		30.0
88	4.2	96.5	163.2	22.8	116.7	181.9	160.1		31.1
90	4.9	99.8	168.4	24.1	120.6	187.4	165.9		32.2
92	5.6	103.1	173.7	25.5	124.5	192.9	171.9		33.3
94	6.4	106.5	179.1	26.8	128.6	198.6	178.0		34.4
96	7.1	110.0	184.6	28.2	132.7	204.3	184.2		35.6
98	7.9	113.5	190.2	29.7	136.9	210.2	190.6		36.7
100	8.8	117.2	195.9	31.2	141.2	216.2	197.2		37.8
102	9.6	120.9	201.8	32.7	145.6	222.3	203.9		38.9
104	10.5	124.6	207.7	34.2	150.1	228.5	210.7		40.0
106	11.3	128.5	213.8	35.8	154.7	234.9	217.8		41.1
108	12.3	132.4	220.0	37.4	159.4	241.3	225.0		42.2
110	13.1	136.4	226.4	39.1	164.1	247.9	232.3		43.3
112	14.2	140.5	232.8	40.8	169.0	254.6	239.8		44.4
114	15.1	144.7	239.4	42.5	173.9	261.5	247.5		45.6
116	16.1	148.9	246.1	44.3	179.0	268.4	255.4		46.7
118	17.2	153.2	252.9	46.1	184.2	275.5	263.5		47.8
120	18.2	157.7	259.9	48.0	189.4	282.7	271.7		48.9
122	19.3	162.2	267.0	49.9	194.8	290.1	280.1		50.0
124	20.5	166.7	274.3	51.9	200.2	297.6	288.7		51.1
126	21.6	171.4	281.6	53.8	205.8	305.2	–		52.2
128	22.8	176.2	289.1	55.9	211.5	312.9	–		53.3
130	24.0	181.0	296.8	58.0	217.2	320.8	–		54.4
132	25.2	185.9	304.6	60.1	223.1	328.9	–		55.6
134	26.5	191.0	312.5	62.3	229.1	337.1	–		56.7
136	27.8	196.1	320.6	64.5	235.2	345.4	–		57.8
138	29.1	201.3	328.9	66.7	241.4	353.9	–		58.9

* ins Hg Vacuum (Gauge pressures are expressed in inches of mercury if they correspond to absolute pressures of less than 1 standard atmosphere.)

N.B – conversion factors to other units of measure include:

kgf/cm^2 = psi × 0.0703
Bar = psi × 0.0689
N/m^2 = psi × 6 895.0
N/m^2 = Pa × 1.00
1 in. hg = 25.4 mm hg
1 Torr = 1.0 mm hg abs.
kgf/cm^2 = inch hg × 3453

Courtesy of ASHRAE and Du Pont

Refrigerant leak detection

Leaks cannot be tolerated in any refrigeration system, and leak detecting equipment must be well maintained and regularly used during maintenance checks as well as installation work. The methods which can be used with specific refrigerants are listed below in increasing order of efficiency.

Sulphur candles
When lit and exposed to air containing ammonia vapour, these give off a white cloud of ammonium chloride or ammonium sulphide. This method cannot be used to pinpoint leaks.

Litmus paper
Moist red litmus paper will turn blue if exposed to ammonia vapour, but cannot be used with any of the halogen family refrigerants.

Bubble tests
Soapy water, a washing up liquid, or better still a purpose-developed leak indicator will indicate the locations of leaks by the formation of bubbles by escaping refrigerant. However, this type of test can only be made on piping or fittings known to be at a higher pressure than that of the atmosphere. Test solutions applied to low temperature, low pressure suction lines could cause considerable damage because the liquid could be drawn into the pipes.

Halide test lamps
Detectors fuelled with propane, butane, or methylated spirits can be used to locate fluorocarbon refrigerant leaks. The detector includes a fuel tank which is, or can be, pressurised to supply fuel at a steady and controlled pressure and a jet to admit the fuel to a burner. When lit, the burner flame is supported by oxygen in the air which is drawn through a tube used as a sensing probe. The probe is passed slowly over the joints or surfaces being leak tested. If any fluorocarbon refrigerants are drawn into the tube, the colour of the lamp flame will change to green or blue, depending on the quantity of gas passed over the burner element.

This type of detector can only be used with non-flammable gases, and care must be taken to avoid igniting any other gases or materials (including pipeline insulation) or damaging heat-sensitive items of equipment. It is also important *not* to fill or pressurise lamps in atmospheres known to contain refrigerant, or to use them when major leaks are known to have occurred. In both cases the sensing element might be contaminated, and give false indications until cleaned or replaced.

Maintenance of this type of lamp starts with keeping it clean, and ensuring that the fuel jet remains clear of obstructions. On lamps using methylated spirits a fine cleaning wire is generally used, either as a factory-fitted device or as a separate cleaning probe. The detector element must be kept clean, and free from scale. Remember that refrigerants help to form contaminated scale which will give false indications of the presence of leaks until the element is removed and cleaned or replaced.

Burner-type detectors are in widespread use, but suffer from one inherent defect. They are not sufficiently sensitive to detect very small leaks which perhaps only allow the loss of an ounce or two of refrigerant a year and which will, if left unrepaired, ultimately result in another service call.

Electronic leak detectors
A wide range of electronic detectors is available, and prices are not prohibitive. All are extremely sensitive – battery operated models for use on site will pick up leaks which give as little as 14 grams (0·5 ounce) per year, and more expensive 'shop or factory' use versions can respond to leaks of as little as 0·014 grams (0·0005 ounce) a year! Clearly, this is the most efficient tool for what can be a difficult and time-consuming job.

The refrigerant is sensed by a plug-in element, exposed to air drawn through a probe or tube. Its presence will be indicated by a flashing lamp, an audible 'bleep' or buzz, or a meter reading, each increasing in speed or intensity as more refrigerant passes over the element.

A special calibration instrument is also available. This enables the more sophisticated detectors to be tested, or set to operate, against a known and precisely controlled rate of leakage of R12. Maintenance of electronic detectors consists of keeping them clean, avoiding knocking or dropping them, maintaining indicator lamps etc, and, in some models replacing the sensing elements when necessary. The use of transistorised and solid state components and circuits make these very reliable items of equipment.

Fig. 4.3 Electronic leak detector
Courtesy of Airserco Manufacturing Company

Leak indicating fluids
Dyes to show up leaks visually are available, and one – du Pont Dytel – is available pre-mixed in service cylinders of R12, R22 and R502. Whilst they work, there are two drawbacks. Some are gradually absorbed by desicants such as activated alumina, often used in liquid-line driers; and they are also likely to damage the moisture indicating medium in combination sight glasses/moisture indicators. These considerations limit the safe use of indicating dyes and the writers recommend that they are not used without reference to manufacturers of equipment.

Refrigeration oils

The lubricants used in refrigeration systems must do more than protect moving compressor parts against wear. In the last event, it is a film of oil which finally seals the suction and discharge valves of a compressor, or the shaft seal of an open type compressor. Oil also acts as a coolant, transferring mechanically generated heat from the *crankcase* to the shell of a compressor. It dampens noise. In hermetic systems, it must not attack the electric insulation. It must remain fluid at low temperatures, and mix well (be miscible) with refrigerants such as R22. It must not contain waxes, or other suspended matter which might clog a capillary tube or the orifice of an expansion valve. Finally, it must remain effective for the life of an hermetically sealed compressor, which can be over 20 years! The choice of oil, and nomination of acceptable alternatives, can only be made by the compressor manufacturers and it is extremely foolish to use other than their recommended lubricants.

Contamination of oil by moisture, or moist air, cannot be tolerated. Oil containers must therefore be stored in dry, well ventilated rooms and not opened until the moment the oil is to be used. Air replacing oil which is being pumped or drawn from its container should be passed through a drier, such as that used in the *liquid line* of the system. Obviously, it is better to use smaller rather than larger containers and to relegate the contents of oil cans which have been used (opened and resealed) a number of times to external use.

It must be remembered that oil is constantly circulated around systems charged with the refrigerants listed earlier (with the exception of R717, which is not readily dissolved in mineral oil). The fluorocarbon refrigerants all dissolve in oil and oil is carried away from the compressor crankcase in refrigerants as they are pumped around the systems. Pipelines and heat exchangers must therefore be designed to help oil flow back to the crankcase. Since the evaporator contents – mainly refrigerant, with some oil – have a higher vapour pressure than those of the crankcase – mainly oil, with some refrigerant – a mixture of oil and refrigerant tends to accumulate in the crankcase during off-cycles. It is possible for the two to separate. This presents the risk of oil being washed from compressor working parts by refrigerant, and of the valves and valve plates being damaged by liquid refrigerant. To prevent this happening, crankcase heaters are fitted to most compressors. These boil off any liquid refrigerant, and ensure that the oil can do its job properly.

The study of lubricants is too complicated to be given more than superficial treatment – although the lubrication arrangements incorporated in various types of compressor, and other practical details, are fully described in the notes on the products concerned – but several points must be remembered.

It is virtually impossible to prevent some moisture from entering oil through contact with air, but vital that oil heavily contaminated with water is not charged into a system. Oil can be tested to see if moisture levels are dangerously high, using the dielectric breakdown voltage method. A potential difference is applied across the oil and the voltage at which the oil breaks down is noted. This voltage decreases with increasing moisture content at a steady temperature (26·7 °C (80 °F) is the standard) and a breakdown voltage of about 25 000 V indicates that no free water is present in the oil. This type of test requires laboratory facilities, and is not one which can be used on site. Here, one can only take every precaution to avoid contaminating oil through contact with air and to ensure that all moisture in a system is removed by thorough evacuation (see Chapter 8). If it is suspected that moisture is present in an operating installation one can test a sample of oil for acid. Under workshop conditions, oil still in original containers and suspected of being contaminated can be dehydrated by blotter pressing or – more realistically – by gently heating the oil and exposing it to a high degree of vacuum, to boil off the moisture content.

Whether the oil used is paraffinic or naphthenic based, or a synthetic alkylbenzene, its individual physical characteristics must suit the refrigerant with which it is used.

The following definitions of oil and characteristics should be memorised.

Entrained oil is that carried, in droplets, by a stream of refrigerant gas.

Miscibility is the ability of a given oil to mix with a given refrigerant, and vice versa. Some oils and refrigerants are attracted to each other more strongly than others.

Viscosity is the resistance of a fluid to flow. It is usually given in the form of the number of seconds it takes a given volume of liquid to flow through a viscosity meter. Scales have varied from country to county and two common ones are:
a) Redwood No. 1 – used in the UK, and expressed in terms of seconds at 37·8 °C (100 °F).
b) Saybolt universal seconds – used in the USA, expressed in similar fashion to Redwood.

These scales can be converted into units of viscosity (pascal-seconds or centipoises) but it is wiser by far to leave them to the compressor manufacturer, and use what he advises.

Pour point is the lowest temperature at which an oil will flow.

Flash or fire points express the boiling point and vapour pressure data of an oil. You might think this unnecessary for oil used in a hermetic refrigeration system, but such data does have to be taken into account when designing systems with high *compression ratios* or other characteristics which might lead to an unsuitable oil being *carbonised*.

Secondary refrigerants

In some large installations it is not possible to use *direct expansion systems* – often because pipes are too long to permit the efficient return of oil being carried with fluorocarbon refrigerants, or have risers too high to permit liquid to be lifted without 'flashing' into a mixture of gas and liquid as the result of drops in pressure. In some cases, particularly where R717 is used, booster pumps are installed to overcome the resistance of long lines; but this is not the safest of refrigerants to have circulating around large, occupied buildings; and secondary refrigerants are frequently used instead.

These can take several forms. In larger air-conditioning jobs, chilled water is the medium most commonly used, being circulated at temperatures ranging between 5·5 °C and 13·3 °C (42 °F and 56 °F). Such temperatures are of no practical value for cold storage or ice-making applications – indeed, one of the things air-conditioning men have to be careful to avoid is accidentally allowing ice to form in evaporators and damaging them. To avoid this a brine solution which remains liquid at temperatures well below – 17·8 °C (0 °F) is then used. If a secondary refrigerant might be exposed to very low ambient temperatures, or there is a need for a non-freezing liquid to be used in outdoor condenser circuits, glycol is often employed.

The following notes summarise the main properties of the most common secondary refrigerants, and comment on applications using them.

Chilled water

This has many desirable features, so long as there is no risk of freezing resulting from exposure to low ambient temperatures. Apart from precautions against freezing, when using chilled water in a sealed circuit we need only add chemicals to inhibit the formation of *scale* in the (usually black steel) pipelines, and install strainers and a make-up/expansion tank to have a safe and stable secondary refrigerant circuit.

The volume of water to be circulated is calculated from the simple formula:

$$\text{Volume} = \frac{\text{Cooling Duty}}{\text{Temperature Rise} \times \text{Mass of Water}}$$

In air-conditioning work, convenient temperature rises through an evaporator cooling and dehumidifying air are in the 4·4 °C to 6·6 °C (8 °F to 12 °F) range. The rise will be slightly more at the water-chiller, due to heat gains through pumps and pipelines, but this is easily allowed for. The only binding temperature considerations when using chilled water are that a temperature below 5·5 °C (42 °F) would require a dangerously low refrigerant temperature in the evaporator, and water entering at a temperature above 13·3 °C (56 °F) would incur unacceptably high suction and condensing temperatures.

Water used to dispose of heat rejected through a condenser poses more problems, since the circuit cannot be sealed if a cooling tower or evaporative condenser is used; and the prevention of scaling and algae growth needs more thorough attention. And we must remember that, as will be explained later in the text, considerably more heat is rejected through a condenser than is removed through the evaporator of a system.

Glycol

It is sometimes not realised that two types of glycol are available – ethene based and propene based. Ethylene glycol (ethane-1, 2-diol) has the better physical properties, especially at low temperatures; but where the secondary refrigerant might come into contact with foods or drink, toxicity considerations may require the use of propylene glycol (propane-1, 2-diol). Both should be inhibited to prevent corrosion of the systems and components through which they are circulated. Their leading physical properties are listed in Fig. 4.4 One of the properties in Fig. 4.4 is specific gravity. This is the mass of a given volume of a substance compared to the mass of the same volume of water, the specific gravity of which is rated as 1·0. Specific gravity is one way of expressing density.

The glycols are normally used as solutions in water and the freezing point of the solution decreases as the percentage by weight of glycol is increased, as shown in Fig. 4.5.

When glycol is used for condenser

Fig. 4.4 Properties of glycols

Property	Ethylene Glycol (Ethane-1, 2-diol)	Propylene Glycol (Propane-1, 2-diol)
Relative molecular weight	62·07	76·10
Specific gravity at 20 °C	1·113	1·036
Boiling point at 760 mm Hg	197	187·4
Vapour pressure at 20 °C (mm Hg)	0·05	0·07
Freezing point	−13·0 °C	−60·0 °C
Specific heat capacity at 20 °C (kJ/kgK)	2·35	2·48
Heat of fusion (kJ/kg) at −13 °C	187	–
Heat of vaporisation at 101 kPa (kJ/kg)	846	688

Fig. 4.5 Freezing points of glycols

% by weight	Freezing point °C (ethylene glycol)	Freezing point °C (propylene glycol)
0	0	0
10	−5	−3
20	−9	−7
30	−16	−13
40	−24	−22
50	−36	−32
60	−57	−57

purposes, a practical maximum temperature which may not be exceeded without risk of the glycol boiling is 82 °C (180 °F) unless the system is a sealed one, in which case 121 °C (250 °F) is permissable. These practices are similar to those used when glycol is used as an anti-freeze in a car's coolant system. Detailed charts must be used to calculate exactly the specific heat capacity and specific gravity of both types of glycol at varying strengths and temperatures. In using the values, remember that the volume formula quoted on page 43 applied to water. For other refrigerants it must be adjusted to make allowance for specific heat capacity. If ethylene glycol at 20 °C (68 °F) were used in place of water, the volume calculated for water would need to be divided by 0·561, which would increase the volume to be circulated by over 78 per cent! Add the effects of the differing specific gravity, and it will be seen that

Fig. 4.6 Calcium chloride brine data

% by weight	specific gravity at 15·5 °C	specific heat capacity at 15·5 °C (kJ/kgK)	Mass (kg) per litre brine solution	$CaCl_2$ in solution	H_2O in solution	specific gravity at 10 °C	0 °C	−10 °C
5	1·044	3·869	1·043	0·052	0·991	1·042	1·043	–
10	1·087	3·580	1·087	0·109	0·978	1·087	1·089	–
15	1·133	3·320	1·132	0·170	0·962	1·134	1·137	1·139
20	1·182	3·086	1·180	0·236	0·944	1·183	1·186	1·190
25	1·233	2·885	1·233	0·308	0·924	–	–	–

glycol circulation rates *must* be properly calculated, not guessed.

Brines

Once more, two types of brine are used: that made with calcium chloride for use in ice plants and ice rinks, and sodium chloride where the brine might come into contact with foodstuffs. For the majority of refrigeration applications, the lower freezing points of calcium chloride make its use commercially popular.

It must not however be forgotten that brines are very corrosive, and *must* be inhibited before use with sufficient sodium chromate to produce an alkaline solution (pH 7·0 to 8·5). The pH level is subsequently adjusted by adding sodium hydroxide (caustic soda) to correct acidity (pH below 7·0) or adding sodium dichromate to correct excessive alkilinity (pH above 8·5). The properties of calcium chloride brine are outlined in Fig. 4.6, but once again a full chart is needed to adequately cover every permutation.

Not the least important point to remember when using chemicals – especially caustic soda – is that 'corrosion' can be felt by the human skin, or eyes, as well as by metals. When handling chemicals, wear goggles and rubber gloves and immediately wash off any which splash onto your skin or clothing. Such items as the brine tanks in ice plants should be drained, examined for corrosion, cleaned, and repainted with an anti-corrosion protector at least once a year. Corrosive effects are speeded up by the presence of air (or, to be more precise, oxygen) and carbon dioxide and wherever possible, systems containing brine should be sealed to prevent exposure to either gas.

Revision questions

1. a) State the nature of heat mostly removed from the refrigerant in an air-cooled condenser and explain how this is accomplished.
   b) If three unmarked gas cylinders containing R12, R22 and R502 respectively are all about half-full and are stored in the same room for several days at a temperature of about 20 °C, explain the method of identifying the cylinder containing the R502.
   c) If 3·99 kJ of heat is added to 50 g of ice at −40 °C, what will the final temperature of the ice be? (Take specific heat of ice as 2·1 kJ/kgK.) (WAEC)

2. a) Explain why oil used in refrigeration systems should not be exposed to the atmosphere.
   b) Explain in detail the two methods of lubrication commonly used in compressors of refrigeration systems. State one advantage of each method. (WAEC)

3. Using the refrigerant pressure/temperature table in the text, quote system pressures corresponding to:
   a) R22 condensing at 40 °C,
   b) R12 suction temperature of −10 °C.

5. Domestic appliances

Refrigerators and freezers

Refrigerators and freezers are most often of vertical construction, with combination models having separate doors to each section (Fig. 5.1(a)). Large freezers in frequent use, ice cream conservators etc., are generally 'chests' – that is, they have a lid on top rather than a door on one side (see Fig. 5.1(b)). This avoids the disadvantage of the 'stand-up' models which, because cold air tends to fall and hot air tends to rise, suffer from spillage of cold air from open doors, and its replacement by hot and often humid ambient air. Moisture from the ambient air which enters the cabinet is deposited as ice on the evaporator or (in the case of freezers) on the contents. Such frost can quickly obscure labels on packages, and it is for this reason that chest-type cabinets are favoured in most supermarkets.

Good practice requires that refrigerators be able to hold fresh foods in the 0–4 °C (32–39 °F) range in ambient temperatures of 43 °C (110 °F), or 38 °C (100 °F) if electricity voltages fall to 90 per cent of rated values. In the same ambient conditions, food freezers should operate between −18 °C and −13 °C (0 °F and 8 °F).

Cabinets

Design and construction must secure an effective compromise between the competing interests of minimal production costs on one hand, and good appearance and operating efficiency on the other. Strength and high insulation values normally result from the use of prepainted steel sheet exteriors, and plastic food storage compartments and inner door liners. These 'skins' are normally bonded together by foamed-in-place urethane insulation – approximately 34 mm (1·35 in) for refrigerators, and 44 mm (1·75 in) for freezers. These thicknesses provide a safety margin of approx 30 per cent above the insulation values necessary to prevent the formation of condensation on the outsides of cabinets operating in an ambient of 32 °C (90 °F) and 75 per cent relative humidity. Good door seals are provided by plastic breaker strips, and magnets which squeeze the door onto soft rubber or vinyl compound *gaskets*.

The weakest point of such a construction is the door gasket, which is exposed to the most wear and tear as the door is opened

a) Twin compartment, vertical model

b) Chest-type freezer

Fig. 5.1 Refrigerator and freezer models

and closed; its magnetic strength is reduced by repeated impacts. Spare gaskets must be stored in the position they will occupy when fitted to a cabinet, to prevent permanent distortion.

Hermetic reciprocating compressors

Nearly all domestic refrigerators and freezers have single-cylinder, welded hermetic compressors and the great majority of these are of the reciprocating type. We must know, and be able to visualise, everything that happens inside a compressor 'can'; so we'll take a good look at how they are made, and how they work.

We will start with the shaft. This is normally machined from a high quality iron casting, or forged from steel. It runs in main bearings – steel-backed, bronze, or aluminium – on either side of the eccentric and connecting rod which transmit the rotation of the shaft to the aluminium or cast iron piston. Figure 5.2 shows the relative positions of the eccentric and the piston.

The piston moves within an accurately machined cylinder which is open at the bottom and sealed at the top by a valve plate. The cylinder block is as simple, and uses as little metal, as possible. It is often aluminium, consisting only of the cylinder, main bearings, and surfaces to locate the block with relation to both the electric motor and the spring-type anti-vibration mountings which secure the assembly within the pressed-steel 'can'.

The valve plate has two types of port – suction ports through which low pressure refrigerant gas is drawn into the cylinder, and discharge ports through which high pressure gas is pumped to the condenser. The valves guarding these ports open and close some 3 000 times a minute, and must be exceptionally strong and responsive. They consist of thin reeds cut from the very best quality steel and are often held against their seats by metal springs. Weakly-tensioned springs then hold the suction valve on the underside of

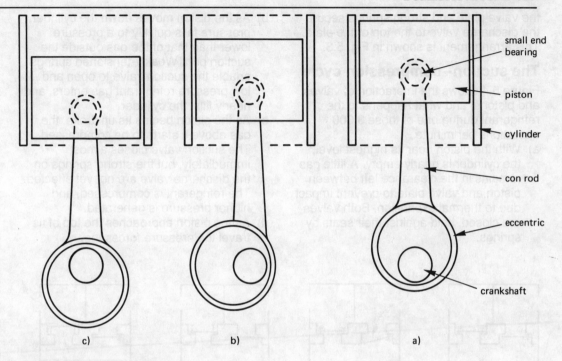

Fig. 5.2 Relative positions of piston and eccentric in a compressor

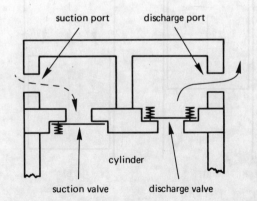

Fig. 5.3 Valve plate layout

TROPICAL REFRIGERATION AND AIR-CONDITIONING

the valve plate and strong springs secure the discharge valve to the top of the plate. This arrangement is shown in Fig. 5.3.

The suction–compression cycle

Figure 5.4 shows the interaction of valves and pistons, and what happens to the refrigerant during one of those 3 000 operations per minute.

a) With the piston near its highest level, the cylinder is nearly empty. A little gas remains in the clearance left between piston and valve plate to prevent impact due to thermal expansion. Both valves are closed, held against their seats by springs.

b) As the piston moves down the cylinder, pressure falls quickly to a pressure lower than that of the gas outside the suction port. Weakly-tensioned springs enable the suction valve to open and low pressure refrigerant gas enters, and nearly fills, the cylinder.

c) As the piston begins its upstroke, the gas above it starts to be compressed. The suction valve closes almost immediately, but the strong springs on the discharge valve are not yet affected. The refrigerant is compressed, and higher pressure is generated.

d) As the piston approaches the top of its travel the pressure forces the discharge valve to open, and the high pressure gas is pumped through it to the condenser.

e) Lastly, with the cylinder nearly empty, pressures over and beneath the discharge valves approach a balance. The valve springs then operate, and close the discharge valve. The system has returned to stage (a) and the cycle of operations starts again.

Compression ratios

It will be seen that the higher the suction pressure, the more gas will be drawn into the cylinder. Refrigeration capacity is higher at higher suction pressures, and lower at lower suction pressures.

Compressor capacity is reduced if the discharge pressure is increased. We have seen that the piston cannot be allowed to touch the valve plate at the top of its upstroke and that some gas is left in the space left for clearance. When the downstroke starts, this residual gas expands, and obstructs the flow of low pressure gas through the suction port. The higher the discharge pressure, the greater the loss of performance since the greater the amount of work that must be done against that pressure.

There is a clear relationship between performance, discharge pressure, and suction pressure. In technical terms, they affect the compression ratio of the compressor. (For a given piece of equipment the maximum performance can be obtained by minimising the compression ratio.)

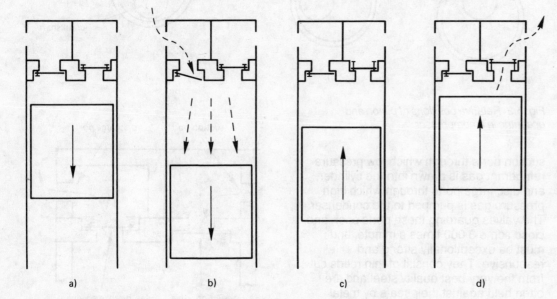

Fig. 5.4 Relationships of piston and suction/discharge valves in a compressor

Compression ratio =
$$\frac{\text{discharge pressure (absolute)}}{\text{suction pressure (absolute)}}$$

Compressor efficiency

The performance of a compressor in terms of the volume of gas it actually pumps, compared with the theoretical volume displaced by the moving piston(s) is called the volumetric efficiency. As the compression ratio of a compressor is increased, its volumetric efficiency is decreased, and a combination of a high compression ratio and low volumetric efficiency requires more power to perform a given amount of work. The measure of power required is called the *performance factor*, and is expressed in terms of power per unit of refrigerating effect. We can confirm these theoretical points by studying a rating table for a typical reciprocating compressor. At a constant suction pressure, if the condensing temperature is increased the compressor's refrigeration capacity will fall, but the power needed to secure that capacity will increase. If the condensing temperature remains unchanged and the suction temperature is reduced, the refrigeration capacity and power drawn will also be reduced; but the power required per TR of refrigerating effect will increase.

Hermetic compressor layout

Returning to our welded hermetic reciprocating compressor, we find that the 'hot gas' discharged from the cylinder passes through a muffler, which reduces

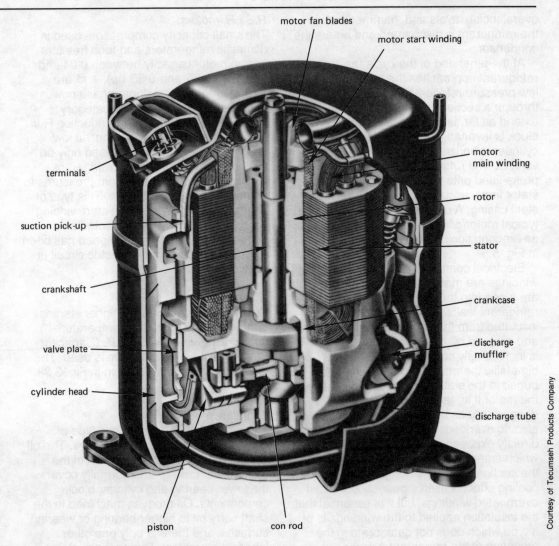

Fig. 5.5 Compressor construction

overall noise levels and 'hammer' before the refrigerant is discharged and enters the condenser.

At the other end of the cycle (as refrigerant approaches the compressor), low pressure refrigerant enters the casing through a suction line which does not extend as far as the cylinder block. The block is invariably arranged so that the cylinder is horizontal, and the shaft vertical. The rotor of the electric motor is press-fitted onto the shaft; and the motor's stator firmly secured within the pressed steel casing. A detailed sectional view of a typical motor/compressor unit – in this case an air-conditioning duty model – is shown in Fig. 5.5.

Electrical connections to the motor windings are made through terminals which are fused into the steel can to prevent refrigerant leaks, the connections being insulated from the can by ceramic casings, and covered by an external terminal box. It is increasingly common practice for a bimetallic thermal overload device to be buried in the stator windings, to disconnect the motor if its temperature rises dangerously as the result of thermal or electric overloads. The motor windings are directly exposed to the cool refrigerant gas which enters the casing and is drawn into the suction port of the compressor. This cooling effect is itself a protection against overheated windings, but it is essential that the insulation applied to the windings is of a type which does not deteriorate in the presence of the refrigerant. Larger compressors often incorporate pressure relief valves to avoid their pumping against excessive head pressures.

R.S.I.R. motors
The small capacity compressors used in domestic refrigerators and food freezers vary in motor capacity between 0·04 and 0·25 kW (0·05 and 0·33 hp), and are designed for use on single-phase power supplies. The motor starting category is R.S.I.R. – Resistance Start, Induction Run (see Chapter 3). It provides only a low starting torque, and can be used only on systems with capillary tube refrigerant controls, which enable system pressures to equalise during the off-cycle. This type of motor has a high resistance start winding which is electrically disconnected by a current relay when running speed has been reached. The schematic electric circuit is shown in Fig. 3.27.

C.S.I.R. motors
Some larger sizes need a higher starting torque when used on low temperature applications, and the C.S.I.R. (Capacitor Start, Induction Run) system is used. The electrical schematic is shown in Fig. 3.28.

Lubrication
The lubrication system for this type of compressor is simple, but effective. The oil is contained in the lower section of the compressor casing, and normally covers the lower bearing and cylinder block components. Oil grooves machined in the shaft carry oil to higher bearing or wearing surfaces, and there is only one other lubrication problem. During the system off-cycle refrigerant will always accumulate in the coldest part of the system – which happens to be the compressor crankcase. R12 and (even more so) R22 mix readily with oil. If no precautions are taken, this can lead to oil being washed out of the compressor when it restarts, or in liquid refrigerant entering the cylinder suction port; in either event, serious damage can result. This is avoided by fitting a low voltage electric heater into or around the compressor oil sump. This operates when the motor is not running, and boils off any liquid refrigerant which tries to enter the oil.

Oil used in compressors operating at low suction pressures can become too hot to be effectively cooled through the walls of the 'can' (a result of a high compression ratio). Such compressors frequently incorporate oil coolers located outside the casing to circulate hot oil through cool ambient air in a suitable heat exchanger, then return it to the sump. Compressors are in fact designed to operate in one of several fairly well defined suction temperature zones shown in Fig. 5.6, and, as explained in relevant sections, motor starting methods for single-phase units are varied to suit the requirements of differing applications.

Hermetic rotary compressors

Several types of small rotary compressors exist as alternatives to the reciprocating type, and the operating principles of two different designs are illustrated in Fig. 5.7.

Rolling piston models
These have the shaft and an eccentric mounted inside a finely-machined roller. This is in the closest possible contact with

DOMESTIC APPLIANCES

Type	Code	Suction range °C	°F
Low back pressure	LBP	−34 to −23	−30 to −10
Medium back pressure	MBP	−23 to −1	−10 to 30
High back pressure	HBP	−7 to 13	20 to 55
Air-conditioning	AC	0 to 13	32 to 55
Heat pump	HP	−26 to 13	−15 to 55

Fig. 5.6 Compressor operating range

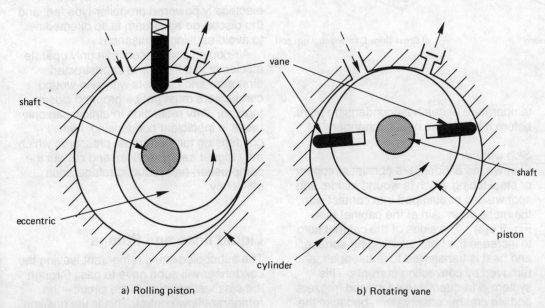

Fig. 5.7 Rotary compressor detail

the wall of the cylinder, and the point of contact runs around the cylinder as shaft and roller rotate. A simple low pressure gas inlet is separated from a reed-type discharge valve by a spring-loaded vane, which remains in constant contact with the roller. When space permits, low pressure gas enters the piston. As the roller rotates, this gas is compressed until its pressure forces open the discharge valve, and the high pressure gas flows out towards the condenser.

Rotating vane compressors

In this design the shaft is set off-centre of the bore, with the circular piston rotating around the inside of the cylinder. Two or more vanes are recessed into slots in the piston, and these are thrown out by centrifugal force as the piston rotates. By maintaining contact with the wall of the cylinder, they effectively divide its free area into sections. The rotation of the piston enables one of these to expand in size whilst another is being compressed. The expanding section fills with low pressure refrigerant from the suction port, and compressed gas is forced out of the contracting area through a discharge valve under high pressure.

It will be seen from Chapter 4 that a special refrigerant, R114, was developed for use in rotary compressors, and that this is a gas with very low pressure characteristics. Both types of compressor have, when operating at low speeds – 1 500 or 3 000 r.p.m. on 50 Hz power – a lower potential compression ability than reciprocating designs, but this can be overcome by substantially increasing the motor, and rotor, speeds. If component clearances are fine, rotary designs suffer little efficiency loss from residual refrigerant; but space has to be left for oil, which provides the ultimate seals between surfaces which are in contact. The strength and durability of the vane, or vanes, is obviously of crucial importance.

Condensers

High pressure refrigerant vapour discharged from the compressors of

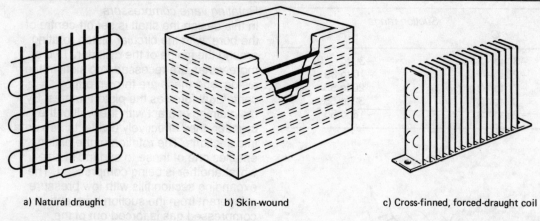

a) Natural draught   b) Skin-wound   c) Cross-finned, forced-draught coil

Fig. 5.8 Air-cooled condensers

domestic appliances is air-cooled to enable the gas to condense. Three types of heat exchanger (condenser is the more usual term) are in common use.

Natural draught
Natural draught types are used on all but the largest domestic refrigerators. They are made from steel tubing arranged in a flat serpentine (snakelike) pattern which is in contact with either steel fins or plates which increase the heat transfer area (see Fig. 5.8(a)). The condenser is fixed vertically to the back of the cabinet. Refrigerant gas enters at the top since it loses heat more rapidly if it is flowing in the opposite direction to cooling air which, taking advantage of convection currents, rises vertically over the hot condenser surface. Condensers are sufficiently large to ensure that after having condensed, liquid refrigerant is subcooled to a temperature beneath its condensing point, before it leaves the heat exchanger.

Skin wound
Skin-wound condensers consist of copper or steel tubing which is wound inside, and spot welded or clamped into contact with, the metal outer skin of the cabinet (see Fig. 5.8(b)). The sides of the cabinet help to increase the heat-exchanging surface; and heat is transferred to ambient air, and removed by convection currents. This system is frequently used in food freezers and ice cream conservators, because the fact that the outer skin of the cabinet is warmed eliminates the risk of condensation.

Forced-draught
Forced-draught condensers, shown in Fig. 5.8(c) are used only on large refrigerators and food freezers, which have a considerable amount of heat to get rid of. They are also used in drinking water coolers, and larger appliances such as air-conditioners. The cooling coil usually has a small surface area, with copper tubes several rows deep. The tube surface area is supplemented by aluminium fins spaced at between 3 and 6 per cm (8 and 16 per inch). Ambient air is blown through this heat exchanger in volumes of up to 110 ltr/s per kJ/s (800 cfm/TR) by an electrically powered propeller-type fan; and the discharge air stream is so directed as to avoid causing a nuisance.

Air-cooled condensers can only operate efficiently in a fresh and unobstructed airstream and cabinets with skin-wound condensers must not be grouped closely together. Any restriction in airflow can only result in inefficient condensing, high condensing temperatures, pressures which may trigger safety devices and cut out the compressor, and reduced refrigeration efficiency.

Liquid lines and fittings

The subcooled liquid refrigerant leaving the condenser will soon have to pass through the smallest opening in the circuit – the refrigerant flow control. This is the position in which any moisture or solid impurities in the refrigerant could most easily cause a blockage and it is therefore common practice to pass the liquid through a fine-mesh strainer, or strainer/drier, immediately it leaves the condenser. In the case of refrigerators and food freezers, the liquid line is often steel tube, which

DOMESTIC APPLIANCES

combines the advantages of low cost and strength.

Capillary tubes

Having been compressed, liquified and subcooled the refrigerant in the 'high side' of the system has been pumped back through the liquid line. It is ready to re-enter the evaporator inside the cabinet, to boil and to remove heat from the cabinet and its contents. Before it can do so the liquid must however pass through some form of refrigerant control, which will have to perform two functions.

a) To meter the *amount* of refrigerant admitted to the evaporator – there must be sufficient to pick up the heat waiting to be removed, but not so much that the evaporator is filled with liquid.

b) To regulate the *pressure* of the refrigerant, and thus help maintain the evaporator at its design temperature.

In the case of the domestic refrigerator, food freezers, and domestic air-conditioner, these duties are usually performed by an economical and efficient control: the capillary tube which is nothing more than a length of fine-bore copper tubing between 3 and 4 metres (8 and 12 feet) long, with internal bore from 0·66 to 2·16 mm (0·026 to 0·085 in). These sizes cover a range of refrigerating capacities from 57 J/s to 5·7 kJ/s (200 to 20 000 Btu/h). The principles of operation of the capillary tube are simple and straightforward. Through it liquid refrigerant is fed from an area of high pressure – the liquid line – into one of low pressure – the evaporator. The walls of the tube offer frictional resistance to the flow of liquid, and refrigerant pressure is reduced as it progresses along the tube. The engineer designing the system knows both the design operating pressures and the amount of liquid refrigerant which must be boiled to remove the heat load from the appliance. He can therefore calculate the amount of resistance to be offered to the flow of a volume of liquid which is related to the volumetric capacity of the compressor, and select a capillary tube with suitable bore and length. In proper balance, this will admit a constant volume of liquid refrigerant to the evaporator. That volume is, however, dependent upon the evaporator pressure remaining fairly steady – in other words, the capillary tube can meet requirements where the refrigeration duty is reasonably steady, as in a domestic refrigerator; but is not able to react to quick and wide fluctuations in the duty which would be required of the evaporator in, for example, a walk-in refrigerator used to freeze sides of beef loaded in daily batches.

It has other advantages, too. When the refrigerator reaches its design temperature and the compressor is cut out by the thermostat, the capillary remains open to both high and low sides of the system. The two pressures 'do what comes naturally' and start to equalise. This reduces the pressure against which the compressor has to be started, and enables a low torque starting system to be used with a relatively inexpensive electric motor.

Each of the appliances described in this chapter use capillary tubes as their refrigerant flow control. If you have to provide reasons for system defects which indicate trouble in the refrigerant control of a domestic refrigerator, or a domestic air-conditioner, do not think in terms of 'expansion valves': domestic appliances almost always use capillary tubes.

Evaporators

Several different types of evaporator (the heat exchanger inside the cabinet) will be found, their design changing with the job they have to do. In domestic refrigerators, the deciding factor generally is the type of defrost system to be used. The three defrost systems listed below are illustrated in Fig. 5.9.

a) Off-cycle defrost models normally use a three-sided 'icemaker' or 'frozen food compartment' manufactured in aluminium plates, metallurgically bonded together after integral refrigerant 'tubes' have been embossed in the metal. The refrigerant channels normally extend onto each surface, and include supply and return *headers*. Alternatively, refrigerant tubes can be *brazed* to the walls of plain metal evaporators. Defrost is normally manual – by pressing a button which ensures that the thermostat does not make circuit for a preset period or by simply switching the unit off.

b) Auto-defrost cabinets need to have some means of disposing of condensate which forms when the coil defrosts. It can be done using a three-sided evaporator and a suitable drip tray piped to a condensate tray

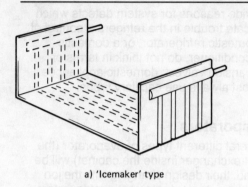
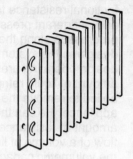

a) 'Icemaker' type   b) Auto-defrost plate   c) 'No-frost' coil

Fig. 5.9 Evaporators

mounted on the back of the cabinet, so that the water is evaporated by the heat of the condenser. Alternatively, a flat embossed plate-type evaporator can be mounted high on the back wall of the cabinet. Condensate is disposed of in the same way, but drainage tray requirements are simplified.

c) 'No frost' models are offered by some manufacturers. The evaporator is usually of finned coil construction, located in an air duct and with air circulation boosted by a small electric fan. Small heaters are sometimes used to remove frost from the coil quickly, during the off-cycle. Condensate drainage is as already described for auto-defrost models.

Shelf-type evaporators

Shelf-type evaporators are used in most upright food freezers, the best designs having a refrigerated surface over the top shelf. Evaporators can be of embossed metal, or serpentine tubing brazed to the underside of flat shelves. Shelves are normally connected in series, and have a suction-line accumulator close to their outlet. The accumulator is simply a shell, with a cross-sectional area much greater than that of the suction line. Any liquid refrigerant which might pass through a very cold evaporator without boiling should expand into this larger area, and change into gas before entering the compressor. (See Fig. 5.10.)

Chest-type evaporators

Chest-type freezers have all sides refrigerated; using either embossed plates, or tubing wound around and tacked onto the inner liner to secure optimum heat transfer. To make best use of the air convection currents inside the cabinet, such coils are wound more closely at the top than at the base of the inner liner.

This type of construction is shared by ice cream conservators and bottle coolers. Mention must be made of the fact that hygienic considerations have long prevented the use of 'wet' bottle coolers, in which bottles are immersed in pre-chilled water. As a consequence, bottle coolers no longer have watertight liners. Anyone trying to use them as 'wet' coolers may find his cabinet leaking, or distorted, or otherwise damaged!

Suction lines

On leaving the evaporator, low pressure refrigerant gas is drawn back to the compressor through the suction line, for the cycle to be restarted. This is a very cool section of tubing, and insulated where necessary to prevent the formation of condensation on the outside of the line. Suction line accumulators are used in the majority of cabinets with frozen food compartments and care is taken throughout the system to prevent the formation of traps which could impede the gravity flow of oil back towards the compressor.

Practical refrigeration circuits

In this chapter we have added some additional components, and a great deal of practical detail, to the basic refrigeration cycle evolved from a study of the fundamental laws of refrigeration, in Chapter 2. Since we cannot afford to forget or to misapply any of these laws, let's work our way through Fig. 5.10 which is a schematic drawing of a practical, working refrigeration system, and make quite sure that we understand exactly *what* is happening in each part of the circuit, and *why*.

DOMESTIC APPLIANCES

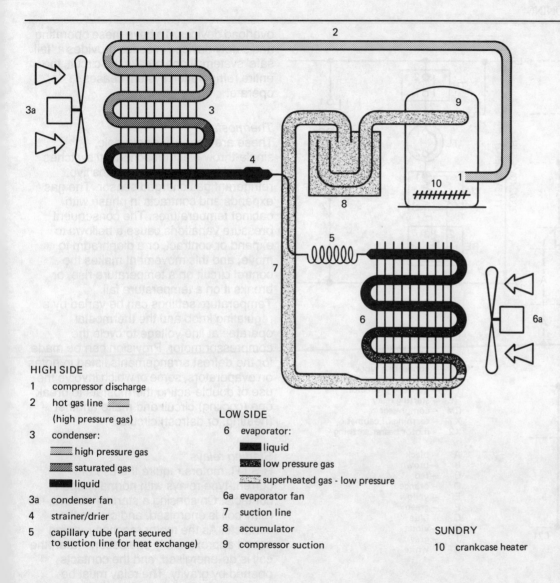

HIGH SIDE
1. compressor discharge
2. hot gas line ▨▨▨ (high pressure gas)
3. condenser:
   ▨▨▨ high pressure gas
   ▨▨▨ saturated gas
   ■■■ liquid
3a. condenser fan
4. strainer/drier
5. capillary tube (part secured to suction line for heat exchange)

LOW SIDE
6. evaporator:
   ■■■ liquid
   ▨▨▨ low pressure gas
   ▨▨▨ superheated gas - low pressure
6a. evaporator fan
7. suction line
8. accumulator
9. compressor suction

SUNDRY
10. crankcase heater

Fig. 5.10 Practical refrigeration circuit

This is one of the most important illustrations in the book, and is applicable not only to domestic appliances, but to most other self-contained, direct expansion refrigeration systems. Cover the condenser and evaporator fans and you have a working circuit for a refrigerator. Uncover the fans and cover the accumulator, and the circuit is typical of a domestic air-conditioner.

Note that in these 'dry' evaporators it is usual to supply liquid refrigerant at the top, and draw superheated vapour from the bottom of the cooling coil. This helps the return of oil to the compressor, and avoids complications which could arise if liquid refrigerant left in a top exit or 'flooded-type' coil continued to boil after the equipment had pulled down to its design temperature, and the compressor been switched off by the system thermostat.

System controls

Reference has been made to the refrigerant control – the capillary tube – and to several electrical operating and safety controls. The type and relationship of these electrically operated items is most easily defined by the schematic wiring diagram in Fig. 5.11.

Note that some items – the cabinet light, operated by a door switch, and the crankcase heater (when one is fitted) – must operate when the thermostat has broken the circuit and the compressor ceased to run. They are therefore wired in parallel with the thermostat and compressor; which in turn are arranged in series with the compressor starting and

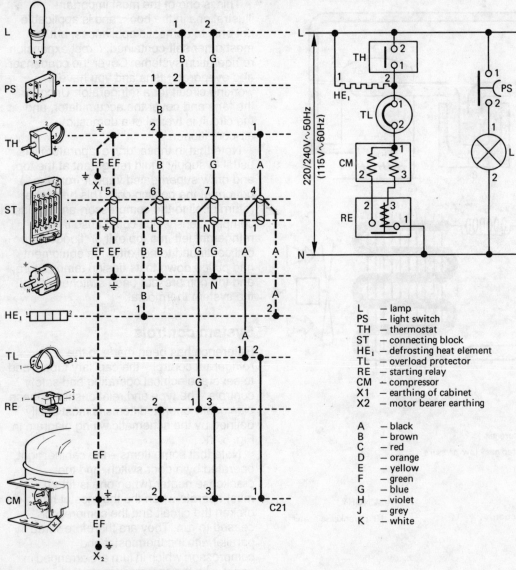

Fig. 5.11 Refrigeration wiring diagram

overload devices. To have these operating and safety devices in series provides a 'fail safe' system: if one breaks the circuit, the entire refrigeration system ceases to operate.

Thermostats
These are normally single-pole, single-throw electromechanical switches actuated by a temperature sensitive, refrigerant gas charged sensor. The gas expands and contracts in phase with cabinet temperatures. The consequent pressure variations cause a bellows to expand or contract, or a diaphragm to move; and this movement makes the control circuit on a temperature rise, or breaks it on a temperature fall. Temperature settings can be varied by a regulating knob and the thermostat operates at line voltage to cycle the compressor motor. Provision can be made for the defrost arrangements listed in notes on evaporators, some of which involve the use of double-acting thermostats to break one (cooling) circuit and make another (heating, or defrost) circuit.

Starting relays
R.S.I.R. motors require the use of current-type relays with normally open contacts. On sensing a starting current, the relay coil is energised, and closes the contacts. As the motor reaches running speed approximately one second later, the coil is de-energised, and the contacts opened by gravity. The relay must be installed with the coil vertical, to enable contacts to make or break cleanly. This

DOMESTIC APPLIANCES

type of relay is used also with C.S.I.R. motors.

Compressor overloads
External, line-break overloads mounted on the compressor casing incorporate bimetallic discs. These react to the thermal effects of electrical overload or overheating by flexing downwards, and breaking the circuit to the compressor motor, as illustrated in Fig. 5.14.

This type of overload is often referred to by a trade name, Klixon. Repeated use may lead to the failure of the disc, the condition of which should always be checked if a compressor refuses to start. In the smallest compressors, this is the sole overload device but in larger sizes the use of bimetallic overloads contained in the motor windings is favoured. The location and construction of such an internal overload is shown in Fig. 5.15.

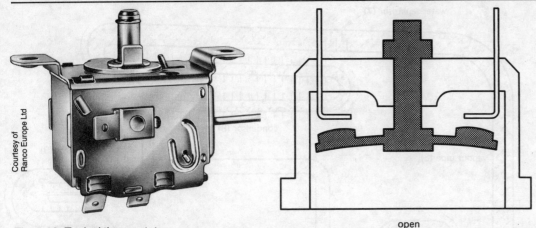

Fig. 5.12 Typical thermostat

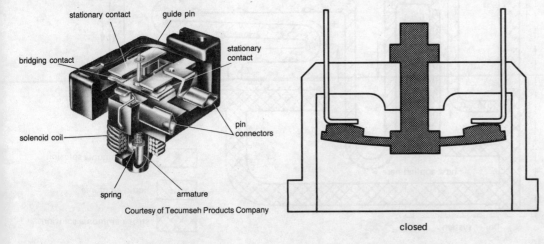

Fig. 5.13 Current-type starting relay

Fig. 5.14 External compressor overload

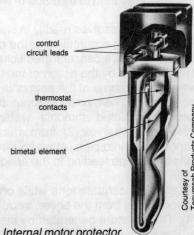

Fig. 5.15 Internal motor protector

Absorption refrigerators (WA)

We have so far examined only mechanical refrigeration circuits. These depend upon the use of electric power, and the compressor, to change the state of refrigerant. This can also be achieved by the absorption system, which requires the use of only low level heat to energise the cycle. Mains or bottled gas, kerosene, or basic electric heating elements can provide the heat, and can therefore supply effective refrigeration in areas not served by mains electricity.

The absorption system is illustrated diagrammatically in Fig. 5.16. It is charged with ammonia, water and hydrogen, and is contained in a sealed circuit at pressures high enough to enable ammonia to condense at room temperatures. The use of ammonia and the high pressure levels require the system to be made of welded steel.

Ammonia dissolves readily in water, and is dense. They accumulate at the bottom of the system, as a concentrated solution. This is diluted by the action of the heater, which 'boils' some of the ammonia at position (1). The mixture of ammonia gas and warm, diluted ammonia–water solution rises into a reservoir (3); from which the vapour rises into pipe (4) whilst the solution falls into a pipe leading to the absorber (10).

Let us concentrate for a while on the vapour rising from the boiler, through point (4). The pressure generated by the heating process forces the vapour up pipe (6) and

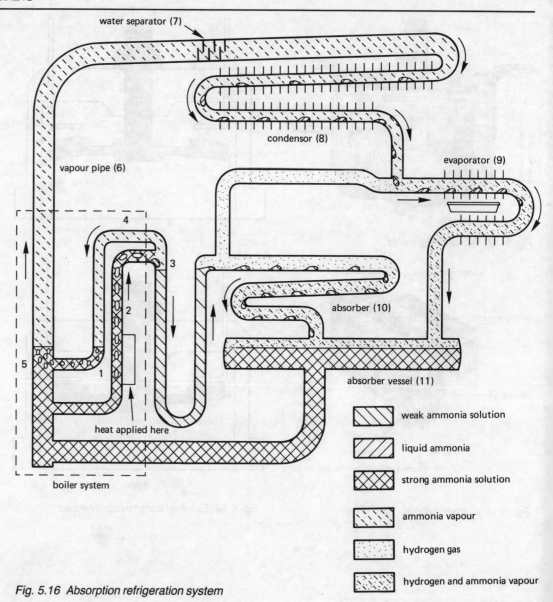

Fig. 5.16 Absorption refrigeration system

through a water separator (7). Here, water vapour condenses, and drains back to the boiler through the vapour pipe. The dry ammonia vapour passes into an air-cooled condenser (8) and resumes its liquid state before flowing into the evaporator (9). The evaporator is connected to the absorber (10), which contains hydrogen as well as ammonia vapour. The hydrogen is a light and lively gas, which seeks always to rise higher in the system. It climbs into the evaporator, creating an area of low pressure which evaporates liquid ammonia entering from the condenser. As usual, the latent heat of evaporation cools the cabinet and its contents. Now mixed, the ammonia and hydrogen gases pass from the evaporator into the absorber vessel (11). This is, as already noted, supplied with a solution of ammonia, the water content of which absorbs the ammonia from the mixture which has left the evaporator. The concentrated solution of ammonia sinks back to the boiler to be recycled, whilst the lighter hydrogen rises towards the entrance to the evaporator. So long as heat is applied to the boiler, the cycle is repeated.

The cabinets of absorption refrigerators – or bottle coolers and frozen food cabinets – appear little different to those used with mechanical systems. The main physical difference is the size and weight of the refrigerant system itself.

Control is thermostatic. It may interrupt the supply to an electric heating element fed from a mains supply, or a battery; or it may interrupt the supply of mains or bottled gas to a burner, the flame being relit by a *pilot light*.

Domestic air-conditioners

The majority of units are either fully self-contained 'room' or 'window' models, designed for installation through either of these structures, or split systems, with an internal evaporator mounted on the floor or a wall, or under the ceiling, plus a condensing unit out of doors. As with domestic refrigerators, competitive pressures have enforced a high degree of similarity in basic designs; but there are, as with other products, significant variations in quality which are not always apparent from selling prices. They are often indicated by comparisons of the surface areas and fin thicknesses of the two air to air heat exchangers, and the durability of condensers exposed to industrially contaminated or saline air.

The self-contained 'room' unit is the cheapest to manufacture and to install, but is invariably more noisy than a split system – usually as the result of compressor noise, and possibly also refrigerant 'gurgle' in the evaporator.

Equipment ratings are quoted against several standards, which should always be clearly stated on manufacturers' literature. The most realistic from the viewpoint of readers in the tropics is the long-standing ASHRAE basis:

Air onto condenser..
35·0 °C DB, 29·4 °C WB
(95 °F DB, 85 °F WB)
Air onto evaporator..
26·7 °C DB, 19·4 °C WB
(80 °F DB, 67 °F WB)

Units for use in very hot climates, or for more temperate areas, should be rated for performance and suitability under appropriate conditions. 'Very hot climates' can be interpreted as 49 °C (120 °F) DB and probably require the use of high efficiency compressors, and possibly also capacitor start and run (C.S.R.) compressor motors.

Models fitted with optional heating arrangements incorporate either electrical heating coils, or reverse cycle valves which interchange the functions of the indoor and outdoor heat exchangers. In either case, a different form of thermostat is needed, the types being examined later in this chapter.

Room units
Cabinets

Basic construction is simple, using pressed steel, and having the baseplate or 'chassis' carried on slide rails which also support the outer 'skin'. This enables the chassis to be slid into or pulled out of the outer cabinet for access to components.

Better quality units are made of steel which has been galvanised, zinc plated or otherwise rust-proofed, and finished in powder-type acrylic paint. The decorative plates (fascias) inside the conditioned space fit flush with, or close to, the wall or window surface, and are invariably in impact-resistant plastic. They carry adjustable air discharge louvres to enable airstreams to be varied in both the vertical and horizontal planes, and a control access panel. The fascia must be easily removable for access to the air filter, and is often

TROPICAL REFRIGERATION AND AIR-CONDITIONING

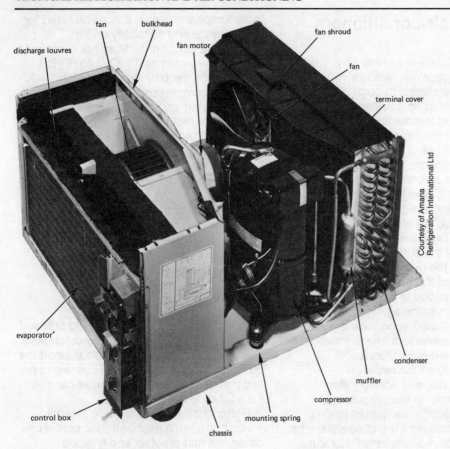

Fig. 5.17 Room air-conditioners

secured by magnetic clips.

It is essential that the two airstreams passing over the evaporator and condenser are effectively isolated. Each has its own fan, both being mounted on a double-shafted electric motor with two or three speed options. The high and low side components are separated by a thermo-acoustically insulated *bulkhead*, through which the fan shaft protrudes to carry the centrifugal-type evaporator fan. This draws room air into the unit through return grille(s) and a cleanable filter, usually made of foamed plastic, before it passes through the cooling coil. Cooled and dehumidified air is then directed into an insulated *plenum chamber*, and discharged into the room through the directional louvres.

Condensate removed from the air drains into the baseplate and flows through a drainage tube to the condenser side of the chassis. This contains the compressor, fan motor, and propeller-type condenser fan in addition to the condenser. Typical component arrangements and airstream directions are shown in Fig. 5.18.

Operating principles – cooling only models

The evaporator temperature is designed to balance room temperature, the volume of room air circulated, and the required sensible heat capacity. Except where air onto evaporator conditions are unusually dry, the evaporator temperature is lower than that of the room air dewpoint, so that some of the air's moisture content condenses on the coil. Typical design temperatures are:

Evaporating temperature	..	7·2 °C (45 °F)
Suction temperature at coil	..	8·3 °C (47 °F)
Room sensible temperature	..	26·7 °C (80 °F)
Room dewpoint temperature	..	15·6 °C (60 °F)
Ambient temperature	..	35·0 °C (95 °F)
Condensing temperature	..	54·4 °C (130 °F)

DOMESTIC APPLIANCES

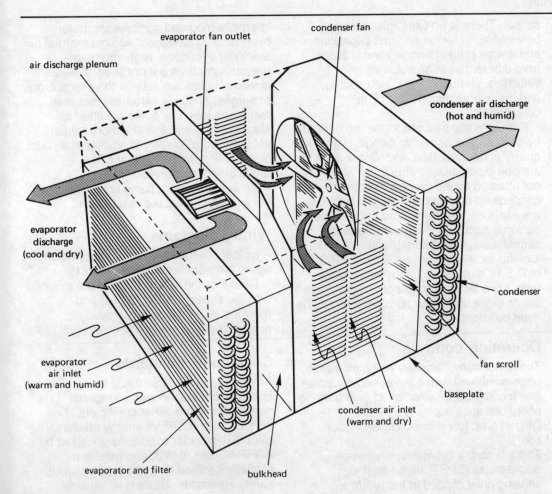

Fig. 5.18 Room unit airflows

Room relative humidity is not controlled, only space temperature being sensed by the controls, but something between 50 to 70 per cent relative humidity normally results, fluctuating with the latent load.

After entering the condenser side of the unit baseplate, condensate removed from the room air is picked up by a slinger ring around the fan blades, and sprayed over the hot condenser. In other than very humid climates, all condensate evaporates on the condenser surface, helping to minimise condensing temperatures and current drawn. In very humid areas condensate accumulates faster than it can be evaporated, and it is then necessary to plumb in a drainage system in plastic or copper tubing. Some units incorporate a removable plug in the baseplate to make this operation as easy as possible. Both evaporator and condenser air velocities must be low enough to avoid the carry-over of moisture, and to maintain acceptable noise levels. The refrigeration system circuit is as shown in Fig. 5.10 (but does not have a suction line accumulator).

Compressors

Hermetic compressor construction is generally as described on pages 47 to 50 and Figs. 5.2/3/4/5, larger sizes including an internal current/thermal overload device and a pressure relief valve. The great majority use single-phase motors, to minimise installation costs. In all cases a crankcase heater operates either continuously or when the compressor is not running. R22 refrigerant is used regardless of ambient temperatures.

Refrigerant control is by capillary tube, enabling P.S.C. motors to be used in all but the most extreme climates, or where persistent low voltage problems are encountered. (See Fig. 3.30.)

When P.S.C. motors are not likely to have sufficient starting torque to overcome the effects of high ambient temperatures and/or low voltages, C.S.R. motors should be used, or starting gear fitted to P.S.C.

equipment. It is also recommended that the additional starting gear be fitted to P.S.C. compressors fitted as replacements following any burnout of the original compressor motor (see Fig. 3.29).

Air-to-air heat exchangers

All condensers are of the forced-draught type, normally having copper tubes mechanically expanded into aluminium fins spaced at approximately 10 per 25 mm (1 in); the same basic construction is used for indoor heat exchangers, when fin spacing can however be reduced to more like 14 per 25 mm. All-aluminium heat exchangers may be encountered, and should be treated with caution – they are less robust than copper and it is extremely difficult to repair high-side leaks resulting from mechanical damage. Fins which are spiral wound or corrugated onto tubes are very difficult to keep clean. Copper finned coils are not financially viable, since where units operate in extremely corrosive ambient air the steel components are in practice found to corrode almost as quickly as good quality aluminium fins. A more realistic alternative is to use acrylic-dipped condensers of standard copper/aluminium construction.

N.B. – it is *not* necessary to provide solar screens for air-conditioner condensers.

Fans

As already stated, two or three fan speeds are normally available, enabling the user to reduce both air volume and noise levels when units are used in bedrooms after sunset. There is no fixed relationship between air volumes and unit capacities, and a wide range of noise levels results from differences in air volumes and velocities, plenum design, insulation quality and the overall dimensions of air conditioners.

Both fans are secured to the motor shaft by grub screws or similar fixings. Higher quality condenser fans and slinger rings are one-piece, plastic structures which are not affected by the properties of the condensate and ambient air to which they are exposed, and cannot be unbalanced by careless handling. Evaporator fans are almost always centrifugal, with forward-inclined impellers. Efficient P.S.C. or C.S.R. motors are used. Many room units have facilities for introducing outside air, or extracting stale room air, through the main bulkhead.

Operating controls

These normally comprise an on/off switch, often combined with a fan speed controller and a compressor selector, to give a range of options such as:
Off; fan only; fan + low cool; fan + high cool.
There is also a bimetallic, single-pole single-throw (SPST) thermostat with sensing phial clipped in the return air stream behind the air filter. The thermostat cycles the compressor, but the evaporator fan runs so long as the unit is switched 'on'.

Safety controls

These comprise either externally- or internally-mounted compressor motor overloads, to safeguard against thermal or electrical overloads. High and/or low pressure controls are *not* fitted. Where power supplies are subject to voltage drops or surges, and only externally mounted overloads (see Fig. 5.14) are fitted as standard, the addition of a more robust voltage-operated relay is advisable, its cost being only a fraction of that of a replacement compressor!

As already noted, some compressors have internal pressure relief valves.

Wiring diagrams

A typical cooling-only room air-conditioner wiring diagram is shown in Fig. 5.19.

Note that the safety controls are in series with the starting relay (if used) and thermostat, providing a 'fail safe' arrangement for the compressor.

Codes and standards

American made units should be certified by independent test houses as regards performance and either coefficient of performance (COP) or energy efficiency ratio (EER) under conditions specified by ASHRAE and AHAM; and most are similarly certified for compliance with UL safety standards. There is increasing pressure on manufacturers for higher standards of efficiency with targets being set by the Department of Energy and 1980 products should have an EER of the order of 7·94 for room units.

$$\text{COP is } \frac{\text{Btu/h cooling capacity}}{\text{watts input} \times 3\cdot412}$$

DOMESTIC APPLIANCES

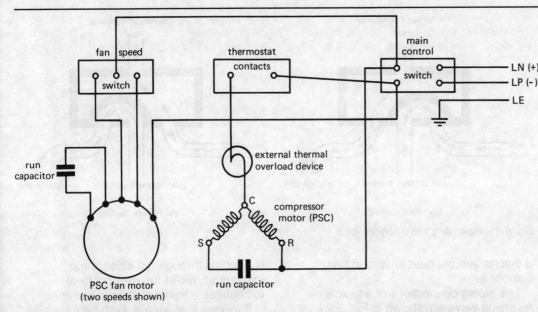

Fig. 5.19 Typical room unit wiring diagram

$$\left\{\frac{\text{Btu/h capacity}}{\text{Btu/h input}}\right\}$$

and EER is $\dfrac{\text{Btu/h cooling capacity}}{\text{watts input}}$

$$\left\{\frac{\text{Btu/h capacity}}{\text{watts input}}\right\}$$

Certification of performance and safety is becoming increasingly sought by manufacturers in the UK and the EEC, and the old practice of rating units by a convenient approximation of the power of the compressor motor, is no longer acceptable.

Since room units are intended only to provide comfort cooling within fairly wide tolerances, they are not often guaranteed to provide specified internal design conditions. Good practice requires, however, that these are within the limits recommended by ASHRAE. In tropical or subtropical areas with ambient temperatures between 32 and 38 °C (90 and 100 °F) DB and a wet bulb of about 27 °C (80 °F), internal conditions of 24·5 to 27 °C (76 to 80 °F) are normally acceptable; and little or no increase in indoor DB temperature is acceptable when using room units in areas with substantially higher outdoor DB temperatures.

Heating unit systems

Heating capability is expected of room air-conditioners in most temperate climates and some much hotter zones where there is a substantial reduction in temperatures after nightfall or in winter. Two forms of heating are available, one very simple and the second more complex.

Electric heating elements

These normally take the form of sheathed, *black-heat* elements of between 1·5 and 3·0 kW capacity (with maximum electrical size limited by the current which can be drawn from ring mains, single-phase power points and conductors) mounted behind the evaporator. Safety precautions are of course essential, to insulate the heating element from the remainder of the cabinet, and to isolate it from mains supply if excessive temperatures build up. This could happen as the result of a thermostat failure or an appreciable reduction in the amount of air passing over the heating coil resulting from such diverse causes as air filters becoming choked, evaporator fans being loose on motor shafts, or unit supply or return air grilles being blocked by external obstructions. In any case, a safety thermostatic cut-out is essential and normally takes the form of a bimetallic strip of simple construction.

Heating unit thermostats

This type of unit must of necessity have a more sophisticated thermostat than the type fitted to cooling only models – one capable of switching over from cooling to heating, or vice versa; and preferably with a differential (a 'neutral zone') say 1 °C wide, to separate the two functions. (The alternative is to rely on a relatively wide

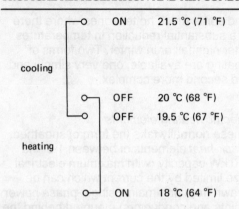

Fig. 5.20 Heating/cooling thermostat switching arrangement

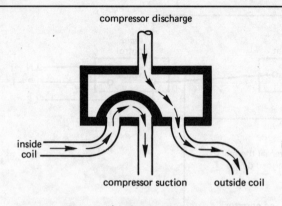

Fig. 5.21 Reversing valve connections

differential to prevent over-reacting to minor variations in the temperature of air at or about set point level). Fig. 5.20 shows a typical switching contact arrangement.

As indicated by the term 'black heat', electric heaters do not increase air temperatures to high levels, the normal range of temperature rise being of the order of 8 to 16 °C (15 to 30 °F).

Reverse-cycle units

The more sophisticated reverse-cycle arrangement introduces a solenoid-operated refrigerant valve, actuated by the thermostat as required, to exchange the functions of what are normally called the evaporator and condenser. It is now better to think of these in different terms – as the 'indoor' and 'outdoor' heat exchangers since, when operating in the heating mode, a reverse-cycle unit uses the indoor heat exchanger as its condenser, through which is discharged the heat absorbed from outdoor air.

The piping connections of a typical reversing valve are shown in Fig. 5.21. In sketch (a) the valve is positioned to operate the air-conditioner in its cooling mode; whilst (b) shows its components positioned for heating duty. The valve's solenoid coil moves a plunger, which opens or closes the ports to two out of three refrigerant lines.

The fourth valve connection – to the hot gas discharge line from the compressor – is not changed, and is physically located on the other side of the *manifold*. With reference to Fig. 5.21(a), in its cooling position the valve directs hot gas to the external (outdoor) heat exchanger (condenser), and channels the low pressure gas leaving the indoor heat exchanger (evaporator) to the suction line. When the valve is reset for heating duties (Fig. 5.21(b)) hot gas from the compressor is directed through the indoor heat exchanger, whilst the external heat exchanger is linked to the suction line.

It follows that, ideally, both heat exchangers should be fitted with refrigerant flow controls which can be bypassed when a coil is not required to act as an evaporator. Additionally, there is in cool climates every possibility that external heat exchangers may frost over and in the majority of models automatic defrosting arrangements are used. The presence of ice can be sensed either thermostatically or from the increased resistance to air passing through the fins of the outside heat exchanger. In the great majority of cases, the defrost control re-activates the reverse-cycle valve, so that hot gas flows through the outside heat exchanger and rapidly defrosts it. The system then reverts to its normal heating cycle arrangement.

It is essential that reversing valves are capable of withstanding and operating against the condensing pressure of the refrigerant in the system, and that they work quietly. Electrically, the solenoid is simple, and thermostat requirements are similar to those outlined for electric heating models.

Heat pumps
There is no clearly expressed and widely accepted definition to indicate the differences if any between reverse-cycle air-conditioners and heat pumps. In point of fact, any refrigeration system removes heat from one place or substance, and rejects it at another, so qualifying for description as a 'heat pump'. Until authoritative bodies produce clear definitions, we suggest that students think 'reverse cycle' when a unit is designed primarily for cooling, but can be operated as a relatively less efficient heating machine, and 'heat pump' when a unit is designed as an efficient heating machine with the capacity to provide a relatively less efficient cooling effect if required. The detailed designs of heat exchangers and condensers, and also compressors used for air-conditioners specifically designed for either of the two main roles differ considerably, and there are limits to the flexibility of any specialised product.

Meanwhile, in cool climates the heat pump is becoming increasingly widely used, since efficient designs are currently capable of producing some 3kW of heating effect for every 1kW of current consumed, so long as the differential between indoor and outdoor air temperatures is not too wide. Neither is air to air heat transfer the only option available; air to water, water to water, and water to air designs are in production. All operate on the same fundamental principles, and use the same general type of reversing valve to change from cooling to heating, or heating to cooling modes. This is not always required and just as we have many cooling-only air-conditioners, an increasing number of heating-only heat pumps is now being manufactured for the European market.

Split system air-conditioners

The basic details of split system units were outlined earlier in this chapter and there is little need to supplement them in respect of

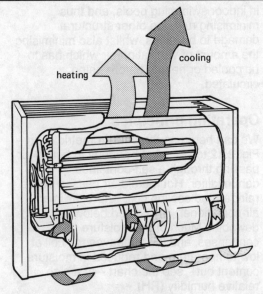
a) Indoor unit

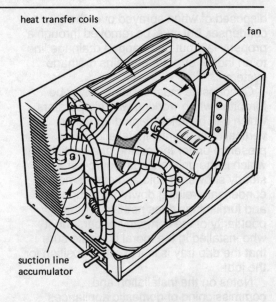
b) Outdoor unit

Fig. 5.22 Split system evaporator

most of their components. The refrigeration system operates on the same principles, and the same electric heating, reverse-cycle heating and heat pump variants are available. It is however necessary to remember a few governing factors which are *not* applicable to room units – the need to supply and fit liquid lines and insulated suction lines between the indoor evaporator and the outdoor condensing unit, and the need also to make positive and permanently plumbed-in arrangements for the disposal of condensate removed by the cooling coil. This cannot always automatically drain back into the condenser base tray, and be disposed of when sprayed over the condenser, but must be removed through a properly laid out and trapped drainage line to an indoor or outdoor mains drainage system.

Condensate must certainly never be forgotton when working with evaporators which can be field-adapted for either vertical or horizontal installation. The presence under the ceiling of a cooling coil which is the unintentional source of a steady shower of cold and often smelly condensate raining down on occupants and furnishings does nothing for the popularity of the company or personnel who installed it; the moral is, make sure that the drip tray is in its proper position for the job!

Notes on the installation and commissioning of domestic appliances appear later in this chapter.

Dehumidifiers

This is a useful item of equipment, designed to maintain the air and contents of rooms used to store documents, spare parts which might otherwise rust, chemicals, etc., at a controlled relative humidity. In the majority of tropical markets, refrigerated dehumidifiers have not been extensively used. This may be due to the fact that units available have been only of very small capacity, and unable effectively to meet commercial-sized duties. This position may change since considerably larger models are being introduced in Europe, where they are of value in reducing the humidity of air in indoor swimming pools, and thus minimising rust and other structural damage to buildings, whilst also minimising the amount of ventilation air which has to be cooled or heated before being circulated.

Operating principles

We can here use the psychrometric chart – Fig. 2.12 to see what happens to humid air passing through a self-contained dehumidifier. Hot, humid air passes over a refrigerated cooling coil, where, as in an air-conditioner, it is cooled below its dewpoint. Much of the moisture condenses, and the air leaves the coil at low temperature and reduced in moisture content but – see the chart – at a high relative humidity (RH).

Here is where the dehumidifier differs radically from the room air-conditioner. The condenser is mounted *inside* the room, and no attempt is made to discharge condenser air to the ambient air. Cold air leaving the evaporator at high relative humidity is passed straight through the warm condenser. The temperature of the air is substantially increased, but since its moisture content is not changed, the temperature rise results in a lower relative humidity – see the chart once more, and trace the full cycle.

Control is generally, but not always, provided, a humidistat and not a thermostat then being used to cycle the refrigeration unit as necessary to maintain room RH in (usually) the 50–60 per cent RH band. The use of an indoor condenser increases room temperature, making the appliance of more practical value in temperate climates than those which require comfort cooling.

A schematic diagram illustrating a cross-section of a typical dehumidifier appears in Fig. 5.23.

Controls

Like a thermostat, the humidistat used operates a single-pole, single-throw switch; to make the circuit on an increase in RH, and break it on a fall in RH to set point level; with a differential of about 5 per cent. The elements now usually take the form of a plastic strip which reacts to humidity changes by expanding or contracting; that movement being transmitted through a

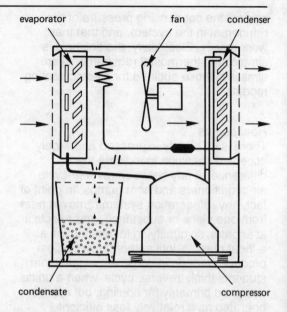

Fig. 5.23 Dehumidifier section

DOMESTIC APPLIANCES

lever mechanism to the electric switch.

Models produced for use in temperate climates may also have defrost controls, similar to those already described for use on the external heat exchangers of heat pumps, to protect the evaporators of dehumidifiers from frosting over if air-onto-coil temperatures are low; but for tropical markets this additional device is best omitted.

Condensate disposal

Much condensate soon accumulates, and the temperate climate practice of supplying a plastic container, stored inside the cabinet and emptied manually when full, is of little practical value in the tropics. The best approach is to use a proper condensate tray with drainage connection, from which condensate is properly plumbed to a drain. If that is difficult because the equipment is installed at a very low level – in a bank vault, perhaps – one has to introduce a galvanised steel sump tank and a condensate pump, with which the condensate can be lifted to drains at a higher level.

Codes and standards

American products should be rated in accordance with ANSI B149·1 (AHAM DH – 1), with performance quoted as the volume of water which they are capable of removing under set conditions:
  26·7 °C (80 °F) DB,
  21 °C (69·6 °F) WB,
  60 per cent RH.
The units must also be capable of operating in higher temperatures:
  32·2 °C (90 °F) DB,
  24 °C (74·8 °F) WB,
  50 per cent RH;
with mains voltages up to 10 per cent lower or higher than rated supply figures. UL Safety Standard 474 should also be complied with.

Installing domestic appliances

Domestic refrigeration

Requirements are simple, and can be summarised in one short list.
a) Check that the electrical characteristics are correct.
b) Wire with the shortest possible cable, into a fused plug.
c) Make sure that the cabinet is level both from side to side and from front to back using a spirit level, to avoid needless noise and vibration.
d) Make sure that no external pipes or fittings are in contact with other metallic surfaces, or the walls.
e) Leave ample room around the cabinet, so that plenty of air is able to rise over the condenser.
f) Do not install cabinets close to cookers, radiators, or other heat generating equipment.
g) If several cabinets are installed together, make sure that hot air from one cannot pass over the condenser of another.
h) Before starting up the equipment remove all packing pieces, slacken off hold-down bolts etc., as detailed in the supplier's instructions.
i) Properly fit any shelves or other items supplied packed in bulk.
j) Do not leave the premises until the cabinet is pulling down to the design operating temperature, is free from noise or vibration and has been tested for correct operation of the thermostat, lights and door switch, defrost control, etc.
k) Make sure that the guarantee card, operating instructions, etc. are handed to the owner, and that he is content with a clean and properly functioning appliance.

Room air-conditioners

Here, installation requirements are more complex, as it is often necessary for a suitable aperture and supporting frame to be fitted in a wall; the unit has to be properly balanced and supported in a manner which prevents noise or 'drumming', regardless of the type of structure in which it is supported; and it may be necessary to plumb in a condensate drain.

Wall apertures
These should be cut to accomodate a timber frame approximately 25 mm (1 in) thick, preferably the same depth as the wall including plaster or other finish and having an internal width and height each approximately 6 mm (0·25 in) larger than the external dimensions of the air-conditioner external casing.

Air-conditioners are normally held in position by two mounting brackets, fixed both to the wall and to the slide rails at the

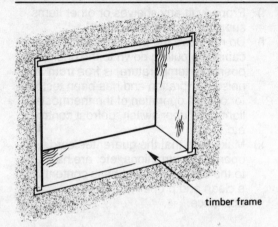

Fig. 5.24 Room air-conditioner wall preparation

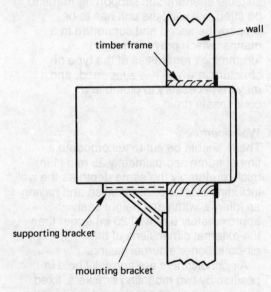

Fig. 5.25 Room air-conditioner support brackets

bottom of the outer casing (see Fig. 5.25). Before proceeding further than securing the frame and brackets, do not forget to make good any damage to the wall – inside and out – around the frame, and to repaint it as necessary. (Some firms make a practice of fitting a decorative surround some 50 mm (2 in) wide around the entire unit.)

When the air-conditioner chassis is to be installed, first remove any packing pieces (most likely to be fitted around fans and fan motor) and test the fan for tightness on the shaft and freedom of rotation without making contact with the shroud or condenser fins. Slacken off any hold-down bolts, in accordance with manufacturer's instructions; check the electrical characteristics and that the unit is free from transit damage, and slide onto the rails in the outer casing. At this point, make sure that the unit is perfectly level from side to side, and that from front to back it is either level or sloping *slightly* towards the rear to help condensate to drain from the front to the rear of the baseplate. If a plumbed-in drainage system is to be provided, now is the time to connect it using a flexible plastic tube to enable the unit to be moved on its runners without pulling out the drain-pipe connections.

The power point serving the unit must be as close as possible to the unit – preferably slightly lower, and to the side on which the power cable enters the front of the chassis – and of the fused switch type.

With the air filter in place, switch on the power supply and start the unit, working steadily through the full operating sequence permitted by the controls. For example work through the sequence: Fan on; Low cool; Mid cool; Max cool; to check that the control switch operates properly in every position which can be selected. Make sure that the thermostat operates, cutting the unit off and restarting it (allowing a two minute interval between the two) before setting it at its central position. The unit should now be beginning to have an effect, with no undue noise or vibration, and a steady flow of cool air from the supply plenum. Push the chassis fully home, securely fix the decorative front panel, and you are nearly through. But not quite! Now is the time to check the full load amperage drawn, preferably using a clip-on ammeter at the unit end of the power cable, and to confirm that this does not exceed values shown on the name plate. Adjust the supply air grilles to produce effective air distribution throughout the room, free from both draughts or pockets of stagnant air. Make sure that you haven't forgotten to close all the windows and doors in the room, and thoroughly clean the fascia and control knobs. Make sure that these are on tight, and are not going to come off in the owner's hand the first time he uses them! Clean up and pack your tools – all except that pocket thermometer. Make sure that the differential between the temperatures of the air entering the return grille, and that leaving the discharge grille, is what the manufacturer says it should be – 11 °C (20 °F) will probably be about right. Then, and only then, are you ready to hand the owner his operating instructions, show him where the controls are, and ask him to sign your job card.

Don't forget that it takes two people to lift

a room unit to waist or chest height and that if it is to be installed at a higher level you will need mechanical aids. The best tool, which can quickly recover its cost, is a high-lift hand truck. This has two forks on which you can sit the chassis at floor level, and a winch with which it can be lifted to any height up to say 3 m (9 ft), and handled in absolute safety.

Split system air-conditioners

Procedures for installing and testing refrigerant lines are fully detailed in Chapter 8, and must be scrupulously followed. From there on, procedures are basically the same as those quoted for room units; although the condensing unit has to be mounted outside, and in high rise buildings it is as well to remember that it must be accessible for service purposes! The best places for condensing units to go are on level concrete plinths with tops at least 150 mm (6 in) above ground level, on a flat concrete roof or on a solid balcony or verandah. Position them so that no nuisance will be caused by the discharge air, and do not forget that power cables must comply with local regulations for conductors and fittings used out of doors.

Dehumidifiers

The main essentials are similar to those of other self-contained appliances – checking levels, electrical characteristics, and the operation of all controls. Care must however be taken to ensure that condensate will be properly disposed of, whether manually or by plumbed-in drains; and that the client is aware of the risk of manually emptied containers overflowing unless they are emptied regularly, at intervals determined by experience on individual jobs. A unit will collect much more than rated water quantities if operated under more humid conditions than those used for performance tests, and it is *not* safe to assume that a container will hold at least 24 hours' condensate accumulation.

Maintaining domestic appliances

Requirements vary by class of product, but one common denominator is the need to keep all plant – including controls and electric power points and cables – clean, ensuring also that nothing interferes with the free supply of air to condensers (and, in the case of room units, from discharge air grilles).

Domestic refrigerators (compressor-type)

This is normally left to the care of their users, whose operating instructions should include the following.
a) Cabinet cleansing – methods to suit the various materials used for exteriors and inner liners are quoted by manufacturers. With many plastics and other materials in current use, no suggestions can be made to cover all products with absolute safety.
b) Defrosting – kitchen knives or other sharp implements must *never* be used, although plastic scrapers are safe on flat surfaces in frozen food compartments. The manufacturers instructions *must* be followed, in view of the wide range of defrosting methods in use.
c) Door gaskets and hinges – care must be taken to avoid damaging doors, either by distorting them or breaking plastic inner liners. Doors of modern cabinets never need to be slammed shut; and heavy handling greatly reduces the life of magnetic gaskets.
d) Condensers – these must be kept free of external obstructions to natural airflow, and should be periodically cleaned with a brush or vacuum cleaner, to maintain optimum condensing efficiency.
e) Upsetting cabinet levels – complaints of noisy operation with little or no reduction in refrigeration efficiency are frequently found to result from a cabinet having been moved, and left standing at an angle.

Domestic refrigerators (absorption-type)

The points listed for compressor models are applicable to absorption types, with the exception that levels are not important in terms of mechanical noise. They do however have a considerable influence on door sealing efficiency, and may affect fuel flow to non-electric models; so the requirement remains a necessary one. In addition:
a) with kerosene-operated units, burners must be kept clean and wicks properly

trimmed, and care must be taken to avoid any damage to lamp glasses, which would affect the proper operation of the heating system;
b) when mains or bottled gas is used burner pilot lights will have to be relit after periods of shutdown, regardless of their cause.

Users should be warned to seek expert advice if the flame will not remain alight, as this may be due to the operation of the fuel safety control.

Domestic air-conditioners

Requirements are substantially different to those for refrigerators. An air-conditioner operates at much higher suction and condensing temperatures (its condenser is in many cases exposed to full sunlight) and is exposed to potentially damaging or corrosive outside influences, including industrially polluted air. For these reasons, professional maintenance is essential if units are to give satisfactory operation for extended periods. The frequency with which maintenance should be carried out depends on local circumstances, and must be left to the common sense of both supplier and user regardless of the other details of contract (which might provide only for labour, or alternatively for labour plus any spare parts which might be required). The main points to watch are as follows.

a) Air filters must be thoroughly cleaned, or renewed, as often as is necessary to prevent the accumulation of dirt; which will quickly reduce the efficiency of the unit.
b) Not less than once a year the evaporator and condenser should be examined, and any dirt or foreign matter building up between the fins must be removed. This *must not* be done with the aid of a steam generator, or a high pressure water jet – we know that it is dangerous to expose refrigerants to high temperatures, or spray water over electrical components. The correct procedure is to use a low pressure spray of purpose-made coil cleaning detergent, which should itself be removed with a *low pressure* water spray, after all electrical components have been removed or screened.
c) Condensate drainage tubes or passages from evaporator to condenser sides of the unit must be checked and kept free from blockages, and unit levels checked and adjusted.
d) At least once a year the condition of all metal components should be checked. Any rust should be removed, and ferrous surfaces repainted with an anti-corrosion preparation.
e) Regular checks should be made to confirm that both fans are tightly secured to the motor shaft by their locking screws and any necessary adjustments made.
f) All operating and safety controls should be tested for correct operation at least once a year, and the condition of electrical contacts and connections should be checked for tightness and freedom from corrosion or pitting.
g) Routine checks should always include air temperatures onto and off the evaporator, freedom from noise and vibration, and running amperages with machines operating normally.
h) If owners report any loss of operating efficiency, the whole refrigeration circuit should be tested for leaks using an electronic leak detector. Make sure that obvious causes such as dirty filters do not mask something more serious.

General

The detailed recommendations for installation and maintenance of refrigerators and air-conditioners cover the slightly different needs of similar appliances, such as frozen food cabinets or refrigerated dehumidifiers. Installation and maintenance instructions issued by the majority of manufacturers err on the side of caution, but some are dangerously brief. *All* of the general principles listed above are important, and should be borne in mind if they are not included in manufacturers' manuals.

Revision questions

1. a) State four desirable features of a bellow-type shaft seal for an open-type refrigerating compressor.
   b) Describe the constructional detail of such a shaft seal.
   c) Describe a procedure for repairing the shaft seal mentioned above if it is found to be leaking.
   (WAEC)

2. Explain briefly the operating principles of an absorption-type refrigerating unit for use in a domestic refrigerator.

6. Commercial and industrial refrigeration

We hope that before they start to read this chapter – which includes detail on much equipment used in air-conditioning as well as refrigeration applications – students will clearly understand the operating principles of the components of domestic appliances, described in Chapter 5. If they do, they will find that in moving on to a similar study of larger and more complex equipment, we do not have to enter any strange new technical territory. The basic refrigeration cycle remains that illustrated in Fig. 5.10. If we understand how and why components of larger plant vary from equipment with which we are fully familiar, we shall be able to work on large installations without worry or doubts. We shall of course have to master some additional theory, but our earlier studies will enable us to proceed in logical steps, continually adding to our store of basic knowledge.

Reach-in refrigerators

Commercial requirements for storage cabinets up to say $0.4 \, m^3$ (15 cu. ft) require the use of several types of refrigerator. Apart from the different temperatures needed by various types of foodstuffs (see Fig. 6.35), the cabinets may have a few large doors, or more smaller ones, to suit the convenience of the user. He may in fact need doors on both sides of a cabinet, to enable trays or containers to be loaded on one side and removed from the other – a 'pass through' design. The doors may be partly or wholly glazed to suit them for use in self-service stores, with two layers of plate glass for medium temperatures, and three for frozen-food cabinets. The window assemblies are perfectly sealed, and all air is evacuated from spaces between panes of glass to increase insulating efficiency and minimise risks of condensation forming. Other applications are as diverse as cabinets to store blood or medical specimens, to those operating at the high humidities required by florists.

Cabinets

Whilst older cabinets were often constructed from, and externally clad in, timber with sheet metal or asbestos-based inner liners, current production methods closely resemble those used for domestic refrigerators. Aluminium plasticised steel, or stainless steel inner and outer skins cover foamed-in-situ expanded polyurethane insulation.

Refrigeration systems

In most cases, condensing units – with welded hermetic compressors and forced-draught, cross-finned air cooled condensers – are self-contained. Some manufacturers locate them at the bottom of cabinets, others prefer the top. As cabinet sizes increase, it is increasingly likely that remote condensing units will be offered. These suit locations where it is inconvenient to discharge large volumes of hot condenser air, or where high standards of hygiene must be observed.

In either case the evaporator will – like the condenser – be of the forced draught type, air usually being discharged at high level after having been drawn upwards through finned cooling coils. These are often mounted upright in the *mullions* between doors, between baffles which form air ducts. Defrosting and condensate disposal options are basically the same as those used in domestic refrigerators. With a few exceptions which will be specifically noted, no absorption systems will be encountered in non-domestic equipment.

Whilst most reach-in models will use R12, some applications require changes in the designs of specific system components. As in Chapter 5, we will examine these in detail before describing complete appliances or installations. The exception made in the case of reach-in refrigerators (it could as well have been refrigerated display cases) was intended only to illustrate the earlier point that, so long as you have mastered the basic theory, you will find little difficulty in

TROPICAL REFRIGERATION AND AIR-CONDITIONING

Fig. 6.1 Reach-in refrigerator

understanding larger systems. Our account was in fact incomplete, since some reach-ins need controls and other components of types not previously described. We will study them all, in the same order used in Chapter 5, to help you to work around different types of system in one logical progression.

Cabinets and insulation

Regardless of usage, cabinet insulation has two main functions: to reduce heat gains to acceptable levels, and to prevent condensation forming on cold external surfaces. If such condensation penetrates the outer skin, it might reduce the efficiency of the insulation, and then cause it or the structure with which it is supported, to lose its strength. In short, insulation must prevent the outer surfaces of cabinets from falling below the dew point temperature of the surrounding air; and the outer skin must be vapour-sealed to prevent any condensation which might form under extreme conditions from penetrating that skin. Our studies of the basic theories of changes of state, and the pressures exerted by air and by vapour, enable us to understand how and when condensation will form.

The foaming of polyurethane insulation between rustproof metal skins prevents moisture from entering the structure under any circumstances so that no further vapour seal is necessary; and since such insulation is strong, no supports are needed between the skins, to hold them in position. This absence of 'heat bridges' enables the structure to offer more resistance to heat flow than did older designs, in which insulation, liners, doors, etc. were supported on timber or metal frames.

Resistance to heat flow has been accurately established for all structural materials, and is referred to as 'R'. The measure of the quantity of heat which can pass through a material, or structure is known as its 'U' factor. This is 1·00 *divided by* 'R'. The higher the resistance, the lower the heat transfer coefficient 'U', which is expressed in terms of units of heat per unit area per degree of temperature difference at each side of the structure. In British Units, 'U' is quoted in $Btu/h.ft^2.°F$; Resistance to heat flow is increased if the thickness of insulation is increased, if an air gap, or a vacuum, is introduced between two layers of material, or if one of the surfaces is lined with a reflector such as aluminium foil, which turns back some of the heat energy attempting to pass through the structure.

Insulating materials vary greatly in efficiency and in their basic properties. Foamed polyurethane is a more efficient insulator than foamed polystyrene, glass fibre or cork – by a factor of approximately two. Only about half the thickness of foamed polyurethane is needed to achieve a required 'R' value than is needed of, say, cork. Another advantage is that it absorbs and transmits less moisture than any of the others. A disadvantage is that poisonous fumes are generated when polyurethane burns. However, this is not a very serious handicap when, as in the case of the majority of refrigerators, it is sealed within fire-resistant skins of sheet metal. As a result, it is used in the majority of domestic and commercial cabinets and factory-made panels for larger structures.

We must note that when 'built-in' insulation is used in a room or building maintained at a low temperature, or a relatively low temperature combined with low relative humidity, it *is* frequently essential that a vapour seal be applied to *all* external structures, when the insulation is applied, unless the insulation possesses an inherent vapour seal. Even then, all *joints* must be vapour-sealed. These considerations apply also when low temperature pipes or ducts are run through areas having a higher dew point temperature; both must be insulated, and vapour-sealed.

Condensing units

The essential components of a condensing unit are:
a) a compressor of welded hermetic, accessible hermetic, or 'open' design;
b) the compressor motor – a stator/rotor unit in hermetic models, or, in the case of 'open' compressors, a drip-proof motor driving directly, or through pulleys and vee belts;
c) a condenser, either air- or water-cooled, with its own fan if air-cooled; and the controls necessary for both operating and safety requirements.

Units designed for use out of doors are enclosed in weatherproof housings, which should be designed to protect electrical items against even heavy rain and they should include an isolator to safeguard personnel working on units out of sight of electric controls and switchgear in indoor plantrooms.

Compressor units are designed for use in indoor plantrooms, with remote condensers. They otherwise have the same components as condensing units described above.

Compressors

Accessible hermetic compressors

Like welded hermetic equivalents, the (generally reciprocating) compressor components and stator/rotor unit are mounted on one drive shaft, totally sealed from the atmosphere. Instead of a welded steel casing, a cast-iron crankcase with bolt-on cylinder head(s) and endplates is used. This construction – most popular in sizes from 5·3 kJ/s to 176 kJ/s (1·5 to 50 TR) for refrigeration work, and 35 kJ/s to 530 kJ/s (10 to 150 TR) for air-conditioning applications – enables motors, valve plates or other components to be removed for repair or replacement. The disadvantages are that castings are heavier and more costly than steel pressings, and occupy more space, and that mechanical components cannot be internally isolated from the casings. Piston speeds are therefore normally lower than those in welded models, to reduce noise and vibration, the usual accessible hermetic unit r.p.m. being that of a four-pole motor – 1 800 on 60 Hz, or 1 500 on 50 Hz supplies, less 'slippage', which gives actual speeds closer to 1 750 and 1 450 r.p.m. respectively. Reduced voltage starting methods are normally star-delta for refrigeration models, and part wind for air-conditioning work. The widely differing suction temperatures and compression ratios called for by the two applications, together with the use of three alternative refrigerants – 12, 22 and 502 – make it difficult to design one compressor which can efficiently satisfy all needs. Most manufacturers concentrate on products suiting the needs of one main type of application.

The internal layout of a typical accessible hermetic assembly is illustrated in Fig. 6.2. Whilst internal components closely resemble those used in welded casings described in Chapter 5, accessible hermetics are invariably fitted with suction and discharge valves incorporating gauge ports. They are therefore more easily and efficiently installed and commissioned than many welded equivalents. Good practice requires the use of both high and low pressure controls (described later) and most compressors have more sophisticated lubrication arrangements. In larger sizes, these include oil pumps mounted on and driven by the shaft. Oil is always continuously filtered, and sometimes further protected by a magnet in the crankcase – to prevent the circulation of metallic *swarf* – before being fed under pressure into the channels leading to bearings and moving components. To prevent seize-ups should the lubricating system fail, larger sizes are fitted with oil pressure safety controls (again described later) and have oil sight glasses and charging ports in their crankcases.

Open compressors

Here, only the compressor components are enclosed in cast-iron crankcase assemblies, the drive shaft projecting through an end cover and a shaft seal which prevents the escape of refrigerant

Fig. 6.2 Accessible hermetic compressor

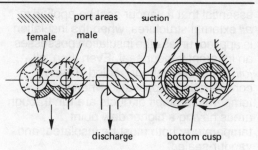

Fig. 6.3 Screw compressor (cross-section)

external drive kit enables compressor speeds to be varied if necessary – normally to a maximum of four-pole motor speed – by changing the motor pulley. Once more, however, there are disadvantages as well as advantages. A self-contained motor cooled by ambient air must be robustly housed, and is more expensive than a stator/rotor unit cooled by low (suction) temperature refrigerant; and we have yet to devise a shaft seal which can be guaranteed never to leak!

The remaining components, and design practices, are similar to those in accessible hermetic models. Note, however, that the absence of internal motor windings makes it practical to use refrigerants which are not compatible with electrical materials. The most important of these is R717; which is extensively used in large, open-design reciprocating and screw compressors used for ice making, cold storage, and food processing applications.

Screw compressors

The helical screw compressor is increasingly used in large systems, and must be examined in some detail. It may eventually replace reciprocating designs for larger refrigeration duties.

The compressor consists of two helically grooved rotors which will interlock. One – the 'male' – has lobes around its perimeter, which fit precisely into gullies machined, screw-fashion, into a matching – 'female' – rotor, as shown in Fig. 6.3.

Low pressure refrigerant entering the suction port enters the space between a lobe and a gulley. As these rotate, the lobe rotates the female rotor and runs

and oil or the entry of outside air. Shaft seals are normally of rotary design. A spring-loaded bellows sits against a 'shoulder' on the shaft, and terminates at the other (outer) end in a finely machined rubbing ring. When the seal is finally assembled, a carbon nose ring is added before the end cover, and held against the rubbing ring by the pressure of the bellows. Both wearing surfaces are perfectly smooth, but the final seal is provided by a thin film of refrigeration oil which separates them, and minimises wear as well as providing the last barrier to equalisation of pressures inside and outside the crankcase. The protruding end of the crankshaft is connected directly, or through a belt and pulley kit, to the driving motor.

Should that motor burn out, there is no contamination of the refrigeration system, and no loss of refrigerant. And the use of an

progressively along the gulley. The gas between the rotors is literally 'screwed' along the length of the gulley, and compressed into a reducing space as rotation continues. Figure 6.4 depicts a section through the length of such a compressor.

There is no direct contact between the male and female rotors, and no mechanical wear other than that on bearings. This ensures smooth and vibration-free operation. As in the case of the shaft seal, the final seal between meshed lobes and gulleys is provided by a film of oil. This is injected at approximately 45 to 50 °C (113 to 122 °F) and has to be cooled to remain within that range. An oil system schematic layout is shown in Fig. 6.5.

Screw designs are suitable for use with any of the main refrigerants – 12, 22, 502, and 717. They can tolerate the occasional entry of small slugs of liquid refrigerant, providing a desirable safety feature on low temperature plant, and they operate at high condensing pressures. There are a number of designs in use, and one currently popular arrangement has a four-lobe male

COMMERCIAL AND INDUSTRIAL REFRIGERATION

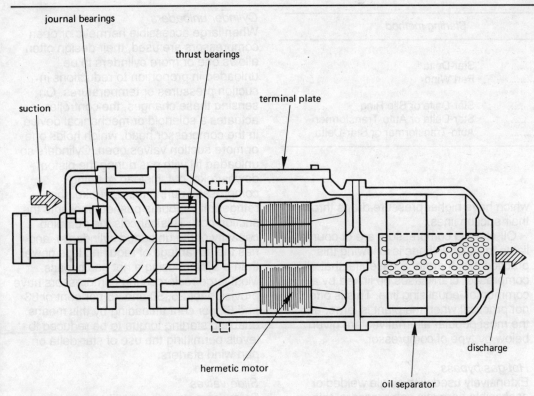

Fig. 6.4 Screw compressor (transverse view)

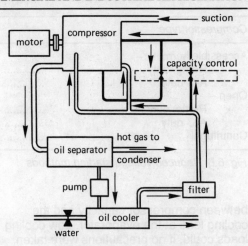

Fig. 6.5 Screw condensing unit layout

rotor turning at two-pole motor speed (3 000 r.p.m. synchronous on 50 Hz) together with a six-gulley female rotor. This consequently rotates at two-thirds of the speed of the male rotor. For each revolution of the drive shaft, such a compressor delivers four measures of compressed refrigerant, through discharge ports which are uncovered as each lobe reaches the end of its travel along a matching gulley.

Rotary Compressors

Some of the types of rotary compressor types described in Chapter 5 are in use and/or under development for use in either open or accessible hermetic systems. We have already looked at their operating principles in some detail, and need not repeat the exercise for commercial models, since the main changes are to size only.

Reduced voltage starting

The principal electrical means of securing reduced voltage motor starting have been described in Chapter 3, but since there are differences between the methods most often used by refrigeration and air-conditioning equipment manufacturers, and resulting also from national practices, the starting methods most often used are summarised in Fig. 6.6 on page 76.

Capacity control and unloading devices

The need to substantially unload compressor cylinders to enable the least complicated – and least expensive – types of reduced voltage starting systems to be used, is one of *two* reasons for the use of capacity control devices. The second is the need to maintain a reasonable balance

75

Compressor type		Starting method
Accessible hermetic –		
Refrig. duty	....	Star-Delta
A-C duty	....	Part Wind
Open –		
Refrig. duty	....	Star-Delta or Slip Ring
A-C duty	....	Star-Delta or Auto-Transformer
Centrifugal	....	Auto-Transformer or Star-Delta

Fig. 6.6 Reduced voltage starting methods

between compressor capacity and the cooling load at the evaporator. Low cooling loads could, if no precautions were taken, cause suction lines to frost back to compressors and the entry of slugs of liquid refrigerant which might damage valves and valve plates, or wash oil from crankcases and lead to mechanical seizures.

Several approaches to this problem are in common use, including – especially when welded hermetic compressors are used – the provision of several refrigeration systems, which can be progressively cycled by a modulating or multi-step thermostat, to balance the cooling load and pumping capacity. This approach has the advantage of keeping power drawn – and hence operating costs – in approximate balance with cooling loads, but it can be expensive in terms of initial cost. If a short-cut is taken, and two or more compressors used with a common cooling coil and refrigeration circuit, precautions have to be taken to prevent any one compressor from accumulating oil at the expense of others which are not running or which have higher pressure drops through their suction lines.

Oil-starved compressors are of course liable to seizure. If one is following this practice it is, therefore, essential that all compressor crankcases be linked by a common oil-equalising line. This is often not practical when the plant is large, and the most popular alternatives are given below by type of compressor.

Hot gas bypass
Extensively used with single welded or accessible hermetic compressors, this approach uses a bypass between the hot gas line and (for best results) a special port on the outlet side of the expansion valve. The bypass is normally sealed by a solenoid valve, which opens if suction pressure falls below a preset value; and a modulating valve then admits into the evaporator a quantity of hot gas which is proportional to the reduction in pressure. The hot gas increases the cooling load and boils off any liquid refrigerant which might cause liquid slugging. This type of valve is fully described in Chapter 8.

Cylinder unloaders
When large accessible hermetic or open compressors are used, their design often allows one or more cylinders to be unloaded in proportion to reductions in suction pressures or temperatures. On sensing these changes, the controller actuates a solenoid or mechanical device in the compressor head, which holds one or more suction valves open. Cylinders so unloaded fill with gas during the piston downstroke, but, instead of being compressed during the upstroke, the refrigerant vapour is pushed back into the suction port. This cancels the pumping capacity of the unloaded cylinder(s), and has the advantage of requiring less power than a system using a hot gas bypass. Most large reciprocating compressors have provision for 25, 50 and 75 per cent *or* 33 and 67 per cent unloading by this means, enabling starting torque to be reduced to levels permitting the use of star-delta or part wind starters.

Slide valves
Screw compressor capacity control can be provided by a slide valve – normally closed – which connects the high and low pressure sides of rotors. It opens in proportion to excessive reductions in suction pressure, and reduces the degree of compression and the refrigeration capacity of the compressor. It can also be used as a means of reducing starting torque.

Compressor safety controls

Mechanical and electrical compressor components, and motors, can be protected by one or more of the following safety devices, which are fitted as standard to all except smaller systems with welded hermetic compressors.

High pressure control

This device, shown in Fig. 6.7, incorporates a bellows connected to (a) a lever mechanism, which operates an electric switch to break circuit if system pressures exceed a preset level, and (b) a capillary tube with its end either brazed or screwed into the compressor discharge valve, or any other circuit position from which the condensing pressure can be accurately sensed. The control setting is made by means of a screw which adjusts the tension of the bellows, the operating pressure being shown on a calibrated scale. Designs are available to provide either automatic reset when system pressures fall below the set point, or hold the switch open until it is manually reset. This second option assumes that the cause of abnormally high pressures will be located and removed before the system is restarted.

Low pressure control

The construction resembles that of a high pressure control, but the switch is piped and wired to break the circuit on a fall below the set point suction pressure and to remake the circuit when the pressure rises above the set point. Its operation is automatic, enabling the control to cycle a compressor in similar fashion to a thermostat.

High and low pressure control

Two elements and switches are contained in a single housing, to cover both functions.

Oil pressure control

This device senses the effective pressure of the oil in a compressor crankcase and operates an electric switch to break the circuit and switch off the compressor motor if oil pressure falls below a safe level. Manual reset is favoured, and in most cases a time delay relay holds the switch closed after a system off-cycle, until the oil pump has had time to build up pressures to normal operating levels.

The effective oil pressure is the difference between the pressure at which it is discharged from the pump, and the pressure within the crankcase itself. The sensing bellows is therefore connected to both locations by capillary tubes linked to both ends of the unit enclosing the bellows (element). The set point is regulated by a tensioning spring, and a lever mechanism operates the switch. Fig. 6.8(a) shows a typical component layout, and Fig. 6.8(b) a schematic wiring diagram for a three-phase installation.

Oil separators

The fact that some refrigerants are not readily miscible with oil, and do not carry around the circuit lubricant which has been pumped out of the compressor, makes it necessary to fit oil separators when they are used, and under certain other conditions also.

Separators are located in the discharge line, and the oil accumulator is piped back to the compressor crankcase, to make certain that oil is returned quickly and positively to the part of the system in which it is needed. R717 systems obviously need separators, and they are used also in large systems using R11, or where R12, 22 or 502 compressor units are installed at a substantially lower lever than condensers, so that oil can accumulate at the bottom of the riser when the compressor is operating partly unloaded (see Fig. 6.9).

A cross-section of an oil separator is shown in Fig. 6.10. Its cross-sectional area is much larger than that of the discharge line. Hot gas entering the separator immediately slows down, so that entrained oil can more easily separate from the refrigerant. This occurs mainly as the gas

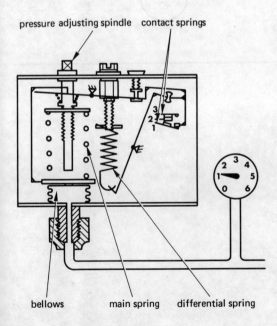

Fig. 6.7 Pressure control

TROPICAL REFRIGERATION AND AIR-CONDITIONING

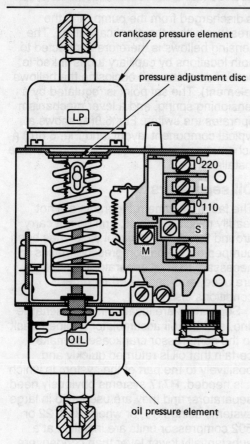

Fig. 6.8 Oil pressure control a) Internal layout

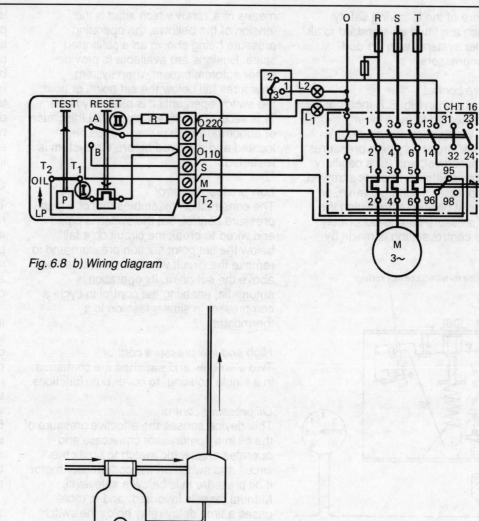

Fig. 6.8 b) Wiring diagram

Fig. 6.9 Oil separator at riser

COMMERCIAL AND INDUSTRIAL REFRIGERATION

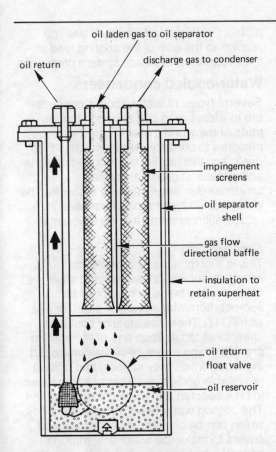

Fig. 6.10 Oil separator cross-section

enable float valves in particular to be serviced.

Separators are normally rated for several refrigerants, at several combinations of suction and condensing temperatures, in terms of the largest system capacity for which each can be used.

Condensers

High pressure, hot gas leaving the compressor must be condensed, and either air or water can be used as the cooling medium.

Air-cooled condensers

These vary little, except in size, from the forced draught types used in domestic air-conditioners. The heat exchanger is still copper tubing, with its surface area extended by aluminium fins spaced at approx 10 per 25 mm. (For R717 units, steel tube and fins are used.) Air is circulated by propeller (or, where noise levels are crucial or condensers have to be installed in indoor plantrooms, centrifugal) fans. Fans are either directly driven by PSC single-phase or three-phase squirrel cage motors, or are belt driven. When condensers are located out of doors they must be contained in weatherproof casings, and suitable precautions taken to protect electrical equipment from rain or damp.

Where equipment operates overnight, particularly in cool climates, to meet loads imposed by processes or equipment rather than external temperatures, there may be a risk of condensing temperatures and pressures falling so low that flash gas can

Fig. 6.11 Air-cooled condensing unit

form in liquid lines. This can disrupt expansion valve operation, or make it difficult to restart a compressor after an off-cycle. Several devices can be used to overcome this problem.

The most effective solution – especially when single-phase fan motors are used – is to vary the speed of the fan motor so that air volumes – and condensing efficiency – are reduced in proportion to low loads at the cooling coil. This is best done by sensing the pressure of the refrigerant in the liquid line, close to the condenser, and transmitting this information to an electronic speed controller which governs fan motor speed. If the liquid-line pressure falls beneath a preset level, the fan speed is reduced and produces a balance between ambient temperature and condensing efficiency.

A similar speed controller senses, not

changes direction, to enter the vertical discharge tube leading back to the hot gas line. Oil accumulates at the bottom of the separator in a reservoir fitted with a float valve. When a suitable volume of oil has accumulated, the valve opens and oil flows back to the compressor crankcase. Access to internal components must be easy, to

79

TROPICAL REFRIGERATION AND AIR-CONDITIONING

the internal pressure, but the external temperature of the liquid line, using a sensor insulated from ambient air, from which operating data is fed back to the speed controller.

An alternative arrangement – which once more has its operation based on unit condensing pressure, sensed through a capillary tube connecting the liquid line and the controller – varies the settings of opposed-blade dampers in the condenser air-discharge outlet. If condensing pressure falls, the dampers close in proportion, and throttle down the volume of air passing over the heat exchanger. Under the extreme conditions of start-up in really cold weather, the dampers would close fully and prevent any air from being circulated. The operating mechanism is a pressure-operated bellows mechanism and is called a bellows-type 'motor' despite the fact that it is not driven by electricity. It is connected by a level to the damper operating linkage, as shown in Fig. 6.12.

Finally, we should take note of the common practice of using several small fans on one condenser, and cycling them thermostatically. A multi-step thermostat reacting to changes in ambient temperatures cuts out fans one after the other as temperature falls, and restarts them in the same sequence as it rises. This method has the advantages of low cost and simplicity, but the disadvantage that ambient air temperatures may bear no relation to the size of the cooling load at the evaporator, or actual system pressures.

Water-cooled condensers

Several types of water-cooled condenser are in widespread use although in many parts of the world there is increasing pressure to conserve water which is fit to drink, and this has led to improved designs and increased use of air-cooled condensers in large capacity systems. The water-cooled types most often used in refrigeration systems are the following.

Shell and tube

This is a compact and economical arrangement with an arc-welded steel shell around tubes made from copper for fluorocarbon refrigerants, or steel for use with R717. There are steel or cast tube sheets welded at each end of the shell to provide a high pressure container around the tubes which they support and bolt-on removable end-covers to enable the tubes to be inspected and mechanically cleaned. The cooling water flows through the tubes, which can be circuited by patterned end covers to make the water flow through several tubes in series before leaving the condenser, if required. Refrigerant vapour enters the shell at a high level, and when it condenses the liquid remains in the lower part of the shell, which acts also as a liquid receiver. The shell is fitted with a liquid-line valve and a pressure relief device, both described later, and may contain baffles to govern the movement of liquid refrigerant. The tubes are often finned to increase their surface area. (See Fig. 6.13.)

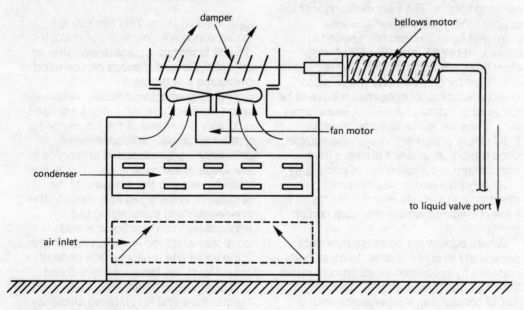

Fig. 6.12 Condenser damper control

COMMERCIAL AND INDUSTRIAL REFRIGERATION

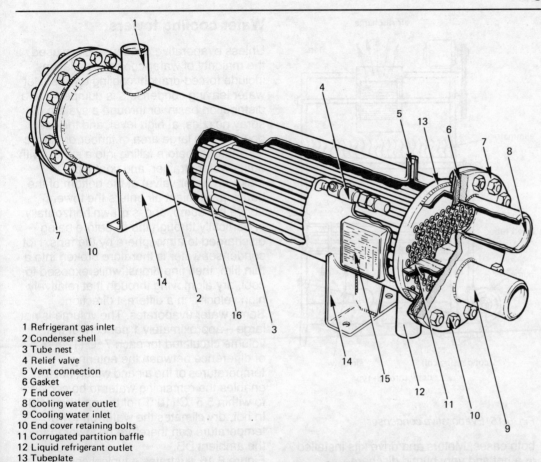

1 Refrigerant gas inlet
2 Condenser shell
3 Tube nest
4 Relief valve
5 Vent connection
6 Gasket
7 End cover
8 Cooling water outlet
9 Cooling water inlet
10 End cover retaining bolts
11 Corrugated partition baffle
12 Liquid refrigerant outlet
13 Tubeplate
14 Mounting feet
15 Design data plate
16 Tube support plate

Fig. 6.13 Shell and tube condenser

Shell and coil
This cheaper construction, much used on smaller capacity units, dispenses with individual tubes and uses a spiral wound tubing coil. Only one end of the shell has a removable cover, at best, and in some cases there are no provisions for internal access. In either case, only chemical descaling is possible.

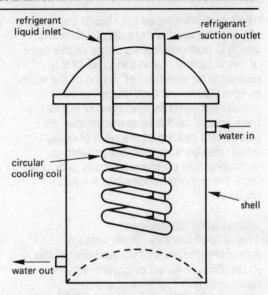

Fig. 6.14 Shell and coil condenser

A liquid-line valve is usually fitted, and safety considerations require a pressure relief valve or rupture disc (see page 83).

Tube in tube
The cheapest construction has a small-bore tube contained within one of larger diameter, both being wound into coils, or looped, to minimise overall size. Cooling water flows through the smaller, internal tube. It is not practical to incorporate pressure relief devices into this type of condenser, and chemical descaling is the only means of removing scale.

Atmospheric condensers
These are normally used only on R717 installations and consist of a number of

81

headered steel pipes erected in the open air, and sprayed with recirculated water which is pumped from a sump at the base of the structure. As in the case of the domestic air-conditioner, some of the water evaporates when it contacts the hot condenser surface, and greatly increases the capacity of the heat exchanger. Hot gas enters the top of the nest of coils, which can be arranged horizontally or vertically, and condensed liquid is drained from the bottom header into a liquid receiver.

Evaporative condensers
These are basically refinements of atmospheric condensers and one of the most efficient types available, whether used with R717 or fluorocarbon refrigerants. The main structural difference is that the (horizontal) coils are encased, under the water sprays, in a galvanised steel or other rust-proof cabinet, through which air is constantly drawn by motorised fans, and then discharged to atmosphere. The constant stream of air, which is relatively cool and dry at the air intake(s), greatly increases the quantity of water which evaporates, and therefore the capacity of the condenser coils. The water is recirculated from a sump fitted with a make-up valve, in the base of the cabinet, as shown in the typical cross-section illustrated in Fig. 6.15.

There are two alternative fan arrangements. One has them above the spray headers, in the leaving air stream; whilst in the other, the fans are mounted at the side of the casings, in the entering airstream, the air discharge being vertical in

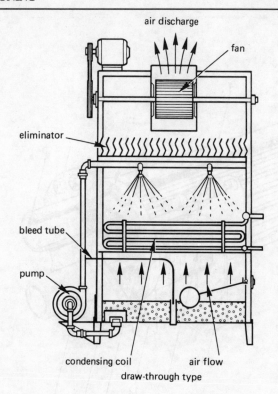

Fig. 6.15 Evaporative condenser

both cases. Motors and drive kits installed in a hot and very humid discharge airstream require more attention than equivalents in a dry, cool position, but both methods provide a high degree of reliability.

Condenser water pumps are invariably factory-fitted, and matched to design pressures and water flow volumes; but liquid receivers and accessories are usually supplied loose for field installation.

Water cooling towers

Unless evaporative condensers are used, the majority of water-cooled installations include forced-draught cooling towers. Hot water leaving condensers is pumped into a distribution basin, or through a system of spray nozzles, at high level, and trickles down over a large area of timber or plastic 'fill' material before falling into a sump (with outlet water strainer, and float-type make-up water valve) at the bottom of the housing. Ambient air enters the towers above the sump, and is drawn horizontally or vertically through the fill before being discharged to atmosphere by the fans. Hot condenser water is therefore broken into a thin film, then fine drops, while exposed to cool, dry air moving through it at relatively high velocity, in a different direction.

Some water evaporates. The volume is not large – approximately 1 per cent of the volume circulated for each 7 °C (12.5 °F) of difference between the entering temperatures of the air and water, but it enables the remaining water to be cooled to within 5·5 °C (10 °F) of the ambient WB. In hot, dry climates the water-off-tower temperature can therefore be lower than the ambient DB.

Figure 6.16 illustrates a typical cooling tower, with the fan at a low level. The remarks on fan positions in evaporative condensers again apply, and it will be seen that the two structures have much in common. Water is however circulated by a remote pump, sized to circulate adequate water volumes against the resistance offered by the condenser and the connecting pipework.

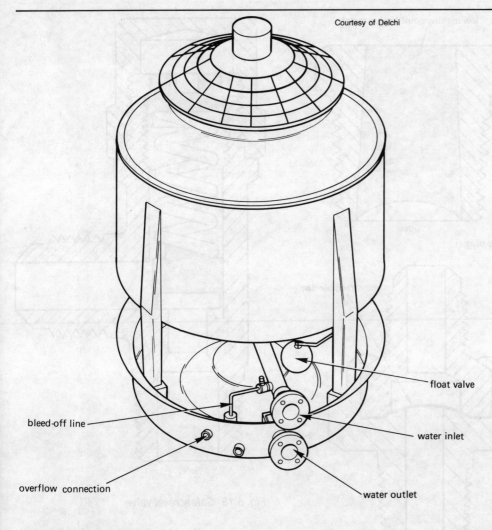

Fig. 6.16 Water cooling tower

COMMERCIAL AND INDUSTRIAL REFRIGERATION

The total water used by a cooling tower exceeds the 1 per cent or so which is evaporated. Not all of the entrained moisture can be removed from air leaving the tower, so some is lost by 'drift' or 'windage'; and more is often continuously bled off from the sump (and replaced through the make-up valve) to inhibit the formation of scale and/or algae. Total losses add up to about 5 per cent of that which would be used if condenser water was run to waste. The majority of designs for tropical areas are based on the temperature of condenser water passing through a tower being reduced by 5·5 °C (10 °F).

Pressure relief devices

It is essential that larger systems incorporate safety devices, to automatically relieve any excessive refrigerant pressures which might be generated. The type used may be specified by local safety codes, but will usually be fitted to the liquid receiver. This is a logical position, since one of the most likely causes of excessive head pressures is the presence of air or other non-condensible gases in the system, and these will always accumulate in the condenser.

Two basic types of safety device are usually described in lectures.

Fusible plugs/rupture discs

A fusible plug is illustrated in Fig. 6.17(a). The fusible metal melts at a known temperature, which corresponds to a refrigerant pressure beneath the safe strength of the receiver, and enables the refrigerant to escape. The same effect can

83

TROPICAL REFRIGERATION AND AIR-CONDITIONING

be secured by the use of a rupture disc (shown in Fig. 6.17(b)) in which the disc will fracture at a known pressure. These are simple and relatively inexpensive devices, but can only be used in small plants since, if they operate, the entire refrigerant charge is lost and the system might later be contaminated by air and/or moisture.

Safety relief valves
The preferred safety device is a pop-type relief valve, illustrated in Fig. 6.18. The valve operating pressure is adjusted by varying the tension of the spring. If system pressures exceed a certain value, they lift the sealing seat. Refrigerant escapes to the atmosphere, but immediately the system pressure falls beneath the set point, the spring re-asserts itself and closes the valve. The refrigerant lost does not exceed the volume necessary to reduce the system pressure to a safe level and there is no risk of contaminants entering the system.

In most countries, large receivers or combination condenser/receivers are termed 'pressure vessels', and must comply with specified standards of construction and performance, to ensure safety and secure the approval of insurance companies. There are many standards in use, which need not be listed, but students should remember that shell-and-tube condensers of horizontal or vertical design are exposed to water as well as refrigerant pressures. It is for this reason that special designs are needed if such condensers are exposed to very high heads of water – as they would be, for

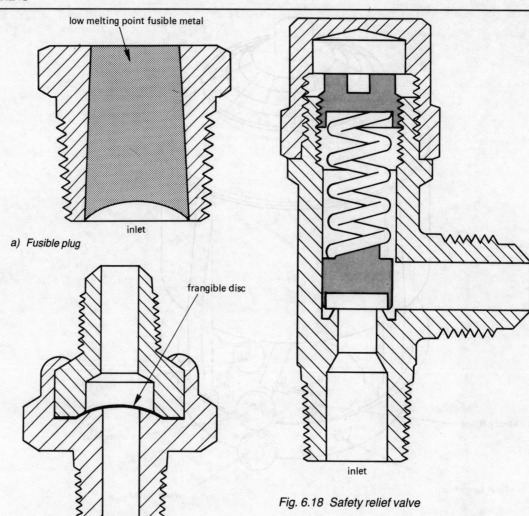

a) *Fusible plug*

b) *Rupture disc*

Fig. 6.17 *Pressure relief devices*

Fig. 6.18 *Safety relief valve*

example, if installed underground in a deep mine and supplied with water cooled by a tower located at ground level.

Condenser capacities

The quantity of heat rejected through the condenser is considerably greater than the cooling capacity at the evaporator. To that capacity is added the heat of compression, mechanically generated heat, and some or all of the heat produced by the compressor motor. Obviously, the last figure is greater if the motor is hermetically sealed and refrigerant-cooled, than if an open compressor with a separate, air-cooled motor is used. And the heat rejected will vary with changes in suction and condensing temperatures, the degree to which the refrigerant has been superheated and the type of compressor – reciprocating or centrifugal.

Condenser requirements are established using data issued by equipment manufacturers, or by authorities such as ARI and ASHRAE. The use of fully detailed tables is not a syllabus requirement, but some typical examples are quoted in Fig. 6.19 so that readers can see how much effect applications have on condenser sizes. They can be expressed in terms of net refrigeration effect (N.R.E.) factors (the percentage of total heat rejected which can be gainfully used at the evaporator) or of evaporator capacity correction factors, applied to the required cooling coil capacity. The two figures are mathematical reciprocals, and our examples are of N.R.E. factors. The evaporator duty is divided by the correction factor to find the capacity required of the condenser.

We must never forget that condensing pressures *must* be kept below the operating limits of the compressors used. In the majority of cases reciprocating compressors are designed for a maximum condensing pressure of about 930–965 kPa (135–140 psi), and it is always desirable to have a safety margin in hand, to avoid the risk of plant cutting out on safety control at the time when it is most needed. Other types of compressor (centrifugal, for example) and absorption systems lose efficiency quickly as condensing temperatures increase, and

Open compressor Suction		Air-cooled or evaporative condenser Condensing °C				
°C	°F	35	41	46	52	57
−40	−40	0·69	0·67	0·64	0·62	–
−29	−20	0·74	0·72	0·70	0·67	–
−18	0	0·79	0·77	0·75	0·73	–
−7	+20	0·84	0·82	0·79	0·77	0·75
+5	+40	0·89	0·86	0·84	0·82	0·80
		95	105	115	125	135
				Condensing °F		

Hermetic Compressor Suction		Air-cooled or evaporative condenser Condensing °C				
°C	°F	35	41	46	52	57
−40	−40	0·53	0·51	0·49	0·46	–
−29	−20	0·62	0·60	0·58	0·53	–
−18	0	0·70	0·67	0·64	0·60	–
−7	+20	0·75	0·72	0·69	0·66	0·62
+5	+40	0·79	0·77	0·74	0·71	0·68
		95	105	115	125	135
				Condensing °F		

Net refrigerating effect factors
Example: Suction 40 °F, Condensing 125 °F, Duty 60 000 btu/h
With hermetic compressor
Condenser capacity = 60 000 ÷ 0·71 = 84 507 btu/h

Fig. 6.19 Condenser N.R.E. factors

need even more careful condenser selections.

Air-cooled models
The capacity of an air-cooled condenser depends upon its surface area, the temperature differential between the refrigerant and ambient air, and the volume of air passing over the coil. (For our present needs, we need not introduce the heat transfer capacities of various types of tube, or fins.) In the majority of cases, air volume will be between 80 and 130 ltr/s per kJ/s (600 and 1 000 cfm per TR) with 110 ltr/s (800 cfm) a fair average. The temperature difference (TD) used on factory-assembled, standard equipment is approximately 16·5 °C (30 °F); but reference to the pressure/temperature relationship table in Fig. 4.2 will show that when air-cooled condensers are used in really hot climates, the TD will have to be greatly reduced to avoid excessive condensing pressures.

Shell and tube models
When packaged equipment is being designed it is usually assumed that recirculated water will enter the condenser 11 °C (20 °F) below the system condensing temperature, and leave the condenser 5·5 °C (10 °F) above its entering temperature, A mathematical calculation will show that this calls for the circulation of 0·19 ltr/s (3·0 US gallons per minute or 2·5 imperial gallons per minute) of water per 3·5 kJ/s (ITR) of refrigerating effect.

As with air-cooled condensers, it is not always possible to work on this basis – wet bulb temperatures may be sufficiently high to make a lower TD, and increased water flow rate, essential. The design engineer has then to ensure that the increased flow rate is acceptable in terms of water velocity and pressure drop through the condenser.

On the rare occasions when water can be run to waste after being used for condenser cooling duties, designers use the assumption that it will enter the condenser 16·5 °C (30 °F) below condensing temperature, and rise by 11 °C (20 °F) before leaving 5·5 °C (10 °F) below the condensing temperature. Here again, water volumes may have to be increased to obtain acceptable results in systems where entering water temperatures are much above 21 °C (70 °F). Oversized condensers must otherwise be substituted.

Evaporative condensers
Assuming that wet bulb temperatures of ambient air permit ready evaporation of water sprayed over the refrigerant cooling coils, evaporative condensers can produce condensing temperatures only about 5·5 °C (10 °F) above the ambient WB. This cannot however be taken for granted, and manufacturers' rating tables must be used to determine air and water needs, and condenser performances.

Condenser water valves

When condenser water is run to waste, or when ambient and/or recirculated water temperatures fall to levels which result in 'over condensing', it is necessary to restrict the amount of water flowing through water-cooled equipment, and a brief examination of suitable control valves is necessary.

In the first case, a two-way throttling valve is used to limit water consumption to that necessary to remove the heat rejected. A sectional view showing the main components and layout of such a valve appears in Fig. 6.20.

A capillary tube allows the condensing pressure to be applied to a spring-loaded bellows, which controls the movement of a stem connected to the valve seat. An increase in pressure forces the stem down, opens the seat, and allows more water to flow through the valve. Decreased

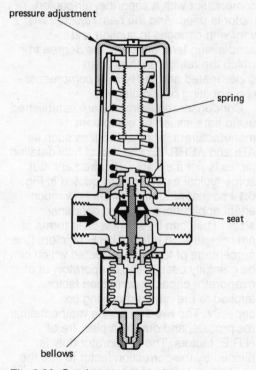

Fig. 6.20 Condenser water valve

condensing pressures allow the spring to dominate, so that the stem is lifted and the valve seat moves progressively towards its closed position. Spring tension can be adjusted to suit the needs of specific installations.

Three-way diverting valves, actuated by changes in pressures or temperatures, are available for use in systems using recirculated water and are particularly useful when one cooling tower and circulating pump serve several condensing units. In cold climates, incidentally, this type of valve may be needed to divert condenser water into indoor storage tanks rather than outdoor sumps, when ambient temperatures fall below freezing point. Such valves are described in Chapter 7.

Condenser design trends

In the recent past, it was unusual to see air-cooled condensers in use with plants much above 175 kJ/s (50 TR) nominal capacity, but this situation has changed radically as the result of demand for water outstripping supply and the increased use of large air-conditioning installations in desert climates where water is a precious commodity. In consequence, the size of air-cooled equipment has greatly increased; factory-assembled packaged units currently reach 1230 kJ/s (350 TR) nominal capacity, and larger plant is field-assembled.

The use of water-cooled equipment, and cooling towers, has in consequence decreased; and will probably become unusual other than in very large installations. Apart from the availability and cost of water as a cooling medium, the reduced maintenance requirements of air-cooled condensers seem likely to assist this trend.

Liquid-line fittings

Valves controlling the flow of liquid refrigerant are described elsewhere; but as we trace our refrigerant around large systems we should comment on the liquid-line sight glasses, strainer/driers and the use of solenoid valves in the majority of liquid lines. First, though, we will consider some facts about liquid receivers, which have been mentioned several times when discussing condensers.

Liquid receivers

Liquid receivers do what their name states – contain liquid refrigerant discharged from the condenser, until it enters the liquid line. Under normal operating conditions they are about half filled with liquid; and a main function is to provide space to hold the full system refrigerant charge if it is necessary to pump down a circuit and, for example, service a compressor. The size of a receiver should enable it to hold more liquid refrigerant than the system charge. In many cases, this capacity is provided by the use of a shell-and-tube or shell-and-coil water-cooled condenser, or an air-cooled condenser with sufficient internal volume to hold the system charge in the liquid, rather than the gaseous, state.

When receivers are used, they normally contain the pressure relief device; and a liquid valve is fitted at the outlet to the liquid line (see Fig. 6.21). Pressure vessel construction standards are desirable.

Courtesy of R.C.A. Receivers Ltd

Fig. 6.21 Liquid receiver

Liquid-line strainer/driers

Strainer/driers are invariably fitted ahead of the liquid-line controls in commercial and industrial equipment. The desiccant charge attracts and removes any moisture present in the refrigerant, until it becomes saturated and has to be replaced or regenerated by heating (preferably in a vacuum) to drive off the moisture from the drying agent. Ths is normally either silica gel, activated alumina or a synthetic molecular sieve, and current practice is to have it moulded into either pellet or cartridge form, to reduce the risk of its being washed out of the drier shell and into the circuit. An additional safeguard is the use of fine-mesh, metal screens on both sides of the desiccant, which catch any solid contaminants as well as holding the desiccant within the drier shell.

As Fig. 6.22 shows, the diameter of strainer/driers is considerably larger than that of the lines in which they are silver soldered, or secured by flared joints. Their increased cross-sectional area reduces refrigerant velocity and increases the exposure of moisture to the desiccant. If

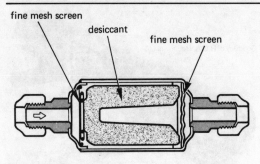

Fig. 6.22 Strainer/drier

driers are regenerated they must be sealed airtight whilst still hot and these seals must not be broken until the driers are re-installed. Larger driers are often made with one end flanged and removable, to enable moulded desiccant charges to be removed and replaced. Driers are normally rated in terms of system capacity and these ratings are a measure of the potential moisture absorbing capacity of the driers.

Sight glasses and moisture indicators

Liquid-line sight glasses are fitted to enable visual checks to be made on the refrigerant charge. If this is low, because of either undercharging or a system leak, bubbles of gaseous refrigerant will be seen in otherwise clear refrigerant liquid flowing through the sight glass. This is normally protected by a screw-on cover and is often double-sided, enabling the refrigerant to be seen clearly if a light is held on the far side of the inspection port.

It is increasingly common practice for factory-charged systems to incorporate a moisture indicator within the sight glass. This is a moisture-sensitive, chemically-treated disc which changes colour progressively if exposed to varying levels of moisture in the refrigerant. A typical indicator is clear green when dry, changes colour to chartreuse – a yellowish green – in the presence of a small amount of moisture, and to clear yellow if moisture content rises to danger level. At that point action must be taken to clean up the system (see Chapter 8).

As already noted in Chapter 4, moisture indicators may be damaged if exposed to dye-based leak indicators. The moral is: follow manufacturers' recommendations or practices.

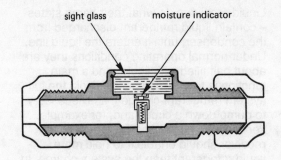

Fig. 6.23 Sight glass/moisture indicator

Solenoid valves

Most non-domestic systems, whether used for refrigeration or air-conditioning, contain liquid-line solenoid valves. These are activated by a thermostat, which causes the valve to shut off the supply of liquid refrigerant to the refrigerant control and evaporator immediately room temperature falls to a set point. All refrigerant downstream of the solenoid valve will then be pumped out of the low pressure side of the system by the compressor before it is itself switched off by the action of the low pressure control. This is called a 'pump-down system' and you must be familiar with the construction, operation and purpose of its components. It is fitted because without it the compressor might 'short cycle' – cut out for a minute or so, then run for an equally short time. This would be the result of small volumes of liquid refrigerant leaking through the expansion valve and re-activating the low pressure control – and would greatly reduce the life of the compressor motor. The solenoid valve ensures that refrigerant is admitted to the evaporator only when the thermostat senses a cooling requirement.

A typical solenoid valve construction is illustrated in Fig. 6.24. Normally, the thermostat breaks the circuit when the system temperature has fallen to set point level, and de-energises the solenoid coil. This allows the plunger to fall, and close the seat of the valve beneath it – liquid-line solenoid valves are normally installed in horizontal pipelines, to help the valve to operate in a positive manner. When the thermostat senses a cooling requirement, its contacts will make the circuit, current will pass through the solenoid and the resultant magnetic force will lift the plunger and re-open the valve seat. Refrigerant will again flow into the evaporator.

The valve illustrated diagramatically in Fig. 6.24 is a simple, two-way type. A number of variations are possible. For

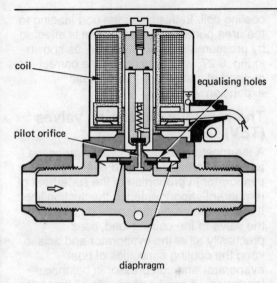

Fig. 6.24 Liquid-line solenoid valve

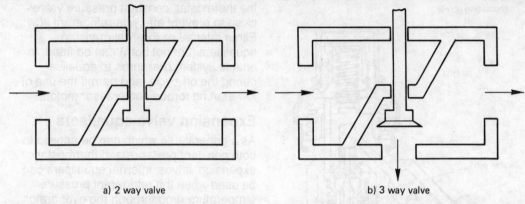

Fig. 6.25 Solenoid valve construction

example, other electrical arrangements can be made to *open* valves on fall in temperature, and internal layouts can be changed to open a second port when the first is closed, as shown diagramatically in Fig. 6.25(b).

The operating differences between these two- and three-way solenoid valves and the reversing type illustrated in Fig. 5.21 are obvious.

Do not forget that whenever a solenoid valve is used in a system, it is *essential* that a strainer or strainer/drier be installed ahead of it. If any contaminants are allowed to enter a valve they might prevent it from opening or closing properly or they might cause it to jam. These events could produce the interesting range of fault symptoms detailed in Chapter 10.

Refrigerant controls

The capillary tubes used in domestic appliances and described in Chapter 5 are seen only on commercial and industrial plant with a relatively steady load, such as ice cream conservators and small reach-in cabinets. Their operating characteristics do not meet the needs of equipment having to react to widely varying loads, or using evaporators which cause a substantial drop in refrigerant pressures. These complications are overcome by the use of expansion valves with 'dry' evaporators, similar to those already described, or float valves to regulate the flow of refrigerant into 'flooded' evaporators which always contain a large quantity of liquid refrigerant.

Constant pressure expansion valves (C.P. valves)

These valves are not very effective in installations having wide variations in cooling loads, and for that reason are not used with large plant. The relationship of the main components is shown in Fig. 6.26.

The fulcrum of this control is the diaphragm, which is subjected to two opposing pressures. Above it, the adjustable tensioning-spring exerts a downward pressure, whilst its lower surface is directly exposed to the pressure inside the evaporator. Since the diaphragm is connected by a push rod to a valve pin which controls the amount of refrigerant flowing through the valve seat, the C.P. valve automatically maintains the evaporator at an even pressure (and temperature). Any increase in suction pressure pushes the diaphragm up, causing the valve pin to move in a closing direction and admit less refrigerant. Any decrease in evaporator pressure results in the set spring pushing the diaphragm downwards, so that the pin moves to a more open position and admits more

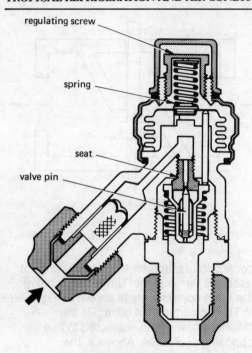

Fig. 6.26 Constant pressure valve (AEV)

refrigerant. The correct set spring setting is one which exactly balances the amount of refrigerant flowing through the valve, and the compressor pumping capacity at design suction pressure.

In practice, C.P. valves should be set to use the entire surface of the cooling coil when the refrigeration load is at a minimum. This prevents refrigerant flooding back to the compressor under low load conditions, or suction pressures rising so high that the compressor is overloaded when the cooling load is increased.

When the compressor is switched off by the thermostat, constant pressure valves close to prevent off-cycle refrigerant flow. Either internal or external pressure equalisers (but not both) can be fitted, to enable system pressures to equalise during the off-cycle, and permit the use of low starting torque compressor motors.

Expansion valve equalisers

As a general rule which can be applied to both constant pressure and thermostatic expansion valves, internal equalisers can be used when the refrigerant pressure/temperature drop through the evaporator does not exceed 1·1 °C (2 °F). In this case, the equalising passage connects the underside of the diaphragm and the discharge port of the valve.

Where the pressure/temperature drop through the evaporator is higher than 1·1 °C, an external equalising device must be used, to avoid excessive superheating of refrigerant vapour before it leaves the cooling coil. In this case, the port leading to the area beneath the diaphragm is affected by pressures in the suction line, as shown in Fig. 6.27, which illustrates the correct method of installing a thermostatic expansion valve.

Thermostatic expansion valves (TEV)

A thermostatic expansion valve varies the amount of refrigerant liquid entering an evaporator in proportion to the superheat of refrigerant vapour entering the suction line. It relates the amount of liquid passed by the valve to the cooling load, uses practically all of the evaporator and acts to keep the cooling capacities of both evaporator and compressor in balance if loads vary. It can be used with all types of direct expansion refrigeration or air-conditioning equipment, regardless of temperatures.

Although the cross-sectional view of a TEV includes both internal and external equalising provision, and both types are available from most manufacturers, the use of external equalisers is necessary on most of the applications in which we are interested. For that reason, we will describe operating principles in terms of externally equalised valves.

As the cross-section suggests, a TEV diaphragm reacts to *three* sources of pressure. That pressing down on the diaphragm is the vapour pressure exerted by a refrigerant charge in a sensing phial firmly secured to a trap-free run of horizontal suction line, as close as possible to the evaporator. When the TEV is in balance the phial pressure equals the sum

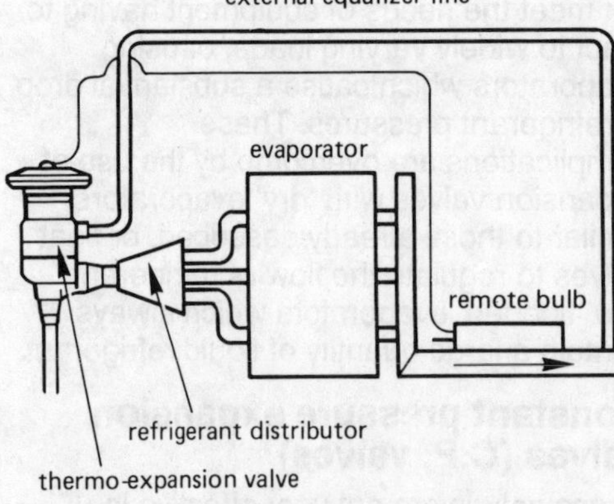

Fig. 6.27 TEV positioning

PIPELINE AND DUCTWORK INSTALLATION

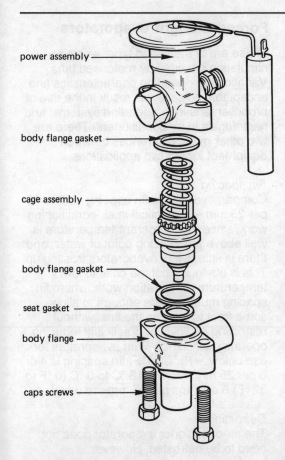

Fig. 6.28 TEV components

of the pressures exerted on the underside of the diaphragm by:
a) the evaporator pressure, sensed through the equaliser, and
b) the adjusting spring, which controls the amount of superheating of refrigerant *vapour* before it leaves the evaporator.

As in the case of the C.P. valve, the pressure acting downwards on the diaphragm seeks to close the valve seat, whilst the opposing upward pressures direct the needle in an opening direction. During off-cycles, the valve is closed by the pressure of the charge in the sensing phial.

When the compressor is running, any increase in the heat load at the evaporator will increase the temperature of the superheated refrigerant vapour leaving it. This will be sensed by the phial, in which the refrigerant charge will expand and exert more pressure on the diaphragm. Moving down, the diaphragm will then open the valve seat and admit more refrigerant to the cooling coil. As the system temperature falls, the sensor charge will contract, and reduce the downward pressure exerted on the diaphragm. This will lift, under the combined evaporator and adjusting spring pressures, and move the valve seat in a closing direction. The amount of liquid admitted to the evaporator is then reduced in proportion to the reduction in the cooling load.

Superheat settings vary with applications and equipment designs, but we should not be too far out if we think in terms of something like 5·5 °C (10 °F) as being typical. Instructions for measuring and adjusting superheat are detailed in Chapter 11.

Expansion valve charges
Before leaving TEV theory, we should note that the sensing-phial charge can be one of several types:
a) **Gas** charges use the same refrigerant as the main system. They exert little pressure on diaphragms. The design helps to limit maximum suction pressures, and to prevent liquid floodback on start-up.
b) **Liquid** charges (sometimes termed cross ambient) have the volumes of the phial, capillary tube and diaphragm chamber so arranged that the bulb will retain some liquid, regardless of external temperature. This charge increases operating superheat as evaporator temperatures fall, and is used on applications with reasonably high suction pressures.
c) **Liquid cross** charges contain a different refrigerant from that used in the main system. This produces different superheat characteristics in low and medium temperature applications.
d) **Gas cross** charges combine the features of gas and liquid cross charges. They provide maximum operating pressures from a small volume of liquid, which is again different from the system charge.

In addition to these types, *adsorption* charges and mechanical pressure limiting means can be used. It is essential that if TEVs are replaced, the new valve must be in all respects identical with the original.

Float valves

Float valves are most often used to control the flow of R717 to flooded evaporators – i.e. cooling coils which always contain a large volume of liquid refrigerant. This is generally the case in brine coolers, the evaporators in ice-maker tanks and some industrial-process liquid coolers. Confusion

TROPICAL REFRIGERATION AND AIR-CONDITIONING

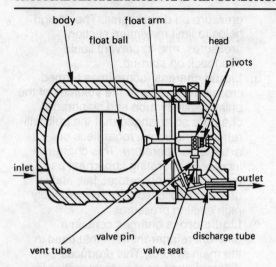

Fig. 6.29 High-side float valve

may arise from the fact that *two* types of float valve are available:

High-side float valves
This type – illustrated in Fig. 6.29 – performs a function similar to that of a TEV; it allows liquid to enter an evaporator at the same rate at which refrigerant vapour enters the suction line. Liquid enters the valve from the condenser, and lifts the float. This movement is transmitted mechanically to the valve pin, which opens the port and allows the liquid to flow into the evaporator. Whilst the mechanism is simple, the refrigerant charge in a system using it is crucial. Too much, and liquid will flood back to the compressor, too little, and system capacity will be much reduced. For practical purposes, the high-side float is therefore restricted to systems with only one evaporator.

Low-side float valves
These perform the same function, but are connected to the low pressure side of the system. When the level of liquid in the evaporator falls below that required, the float opens the valve and more refrigerant flows in from the liquid line. When the liquid in the coil reaches its correct level, the float closes the valve.

When using a low-side float valve the refrigerant charge is not critical and applications include systems having several evaporators supplied from a single liquid header. Each evaporator is of course fitted with its own float valve. This can be mounted in the evaporator, in the surge drum or in an external chamber connected to the evaporator by gas and liquid refrigerant equalising lines.

Provision must be made to prevent excess liquid entering float-fed evaporators during compressor off-cycles, and flooding back on start-up. A liquid-line solenoid valve is the most popular answer. Incidentally, the use of manually operated refrigerant controls is now rare.

Evaporators
Our notes on refrigerant controls mentioned both flooded and dry expansion evaporators. Some dry expansion types were described in Chapter 5, and they are used in the majority of refrigeration and air-conditioning applications in which the evaporators cool air flowing over their external surfaces. Several refinements are however necessary on equipment used in larger systems.

Forced-draught evaporators
These are of the finned type, with air circulation provided by motorised fans. Variations in operating characteristics and acceptable noise levels result in the use of propeller fans in refrigeration systems, and centrifugals in air-conditioners. There are two other major differences between equipment for the two applications.

Fin spacing
Comparatively close fin spacing – 12 to 15 per 25 mm – is practical in air-conditioning work, since the refrigerant temperature is well above the freezing point of water, and there is little risk of evaporators frosting up. This is obviously not the case in low temperature refrigeration work, where fin spacing must be wide enough to allow some frost to form on the fins, without restricting air flows or drastically reducing operating efficiency. With evaporators for use below −18 °C (0 °F) fin spacing is 4 or 6 per 25 mm. From −18 °C to 0 °C (0 °F to 32 °F) 6 or 8 fins per 25 mm are usual.

Defrosting
The air-conditioner evaporator does not need to be defrosted. However low-temperature cooling coils in refrigerators accumulate frost from moisture given off by foods or personnel, or from air entering the space when doors are opened.

The off-cycle defrost methods used in many domestic refrigerators can only be applied to commercial plant operating at well above 0 °C. (In very humid climates a lower limit of 6 °C (42 °F) is advisable.)

Faster and more efficient methods are needed when forced-draught coolers are used in coldstores, freezing tunnels etc. which must be held at low temperatures at all times. There are a number of possible approaches, but for practical purposes only two are important.

Electric defrosting cycles are normally initiated and terminated by time clocks, set to operate as often as necessary to maintain coil efficiency at acceptable levels. Let us take as a typical example the defrost sequence for an evaporator.

a) The time clock breaks the circuit to the fan motor(s) and thermostat. The liquid-line solenoid valve closes, and the compressor pumps down the system and cuts out on low pressure control.

b) Metal-sheathed electric heating elements in contact with the cooling coil, the condensate tray and the drain line are switched on and defrost the coil.

c) When the coil surface temperature rises above freezing point, a defrost-terminating thermostat switches off the heaters, and actuates a time delay relay.

d) After allowing time for all moisture to drain from the coil, the relay re-energises the main thermostat and the fan(s). The solenoid valve opens, and the compressor restarts.

e) The circuit incorporates a safety thermostat to switch off the heating elements if temperatures reach too high a level.

The practical alternative to electric defrost is **hot gas**. A solenoid valve is used to divert discharge gas through the cooling coil for as long as is necessary to secure positive defrost. Either three- or four-way valves can be used, in single or headered, multiple systems; and the refrigerant reversing process illustrated in Fig. 5.21 can be taken as typical. This can be initiated by a time clock as described above, thermostatically or by the reaction of a sensor to thermal changes or to increased resistance to air flow through the evaporator fins. As these principles have been described in Chapter 5, as applied to reverse cycle air-conditioners, details need not be repeated.

In older installations, water was often sprayed over the coils of medium temperature plant as a defrosting medium; or, in low temperature installations, brine sprays served the same purpose. Neither method is much used in modern practice, not only because it is messy and brine might damage foodstuffs, but because the practice is not compatible with the use of cross-finned forced-draught cooling coils. Its decline in popularity has in fact accompanied the disappearance of bare pipe coils for air cooling duties.

Shell and tube evaporators

These are constructed in the same way as the shell and tube condensers described earlier in this chapter, with the refrigerant contained in the shell and the liquid to be cooled pumped through the tubes.

There can be no question of defrost arrangements in this case! If the liquid freezes, it will almost certainly damage the tubes and contaminate the refrigeration system. Precautions to prevent freeze-up are described in Chapter 7, in the section on water chillers. A reminder, though: R717 attacks copper or copper alloys, and chillers using that refrigerant are of all-steel construction.

Submerged evaporators

When brine is cooled in tanks which also contain ice cans, the refrigerant is often pumped through headered steel coils running between or beside the lines of cans, totally submerged to avoid having the brine weakened by condensate. Figure 6.30 shows a good example of a flooded

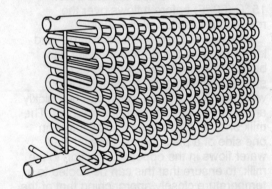

Fig. 6.30 Ice tank evaporator

evaporator (because it contains much liquid refrigerant, not because it is below the level of the brine!) which is usually fed with refrigerant by a float valve.

Baudelot coolers

This type of evaporator is now little used, except in small milk or ice-cream-mix cooling applications. The single-row refrigerant coil is contained between two closely fitting, corrugated stainless-steel outer panels and the whole assembly is usually housed inside removable outer panels to avoid its being exposed to dust or other contaminants. Milk, or other liquid, is fed over the surfaces of the cooler from a distribution trough immediately above it; it is cooled as it descends and is collected in a trough beneath the cooling surfaces. To avoid exposing direct expansion systems to excessively hot liquids, pasteurised milk or ice cream mix passes over two separate heat exchangers. The upper section is cooled by mains or chilled water, which reduces the milk temperature to below 15 °C (60 °F) before it flows over the refrigerated section of the cooler.

With milk being processed in larger and larger centres, the Baudelot cooler has been largely replaced by systems of stainless-steel plates, assembled into watertight assemblies which can be quickly and simply stripped down for cleaning. The milk or other liquid to be cooled stays on one side of a plate; on the other chilled water flows in the opposite direction to the milk, to ensure that this can be cooled to a temperature closely approaching that of the entering coolant.

Where hygiene is not critical, shell-and-tube or shell-and-coil coolers are invariably used for liquid cooling duties.

Bare pipe air coolers

Steel pipe evaporators, coiled on walls and under ceilings, were once used in cold storage or freezing rooms, but no longer have any application. Apart from the limited cooling capacity of unfinned tubes, their great weight and the fact that they occupied much space, these coils were impossible to defrost effectively without removing the produce stored in the coldroom; and they often disappeared under layers of ice which threatened the building structure. The only common current application of bare-pipe coils is liquid cooling for ice making, with coils submerged to take advantage of the fact that water or brine are better conductors of heat than air. This advantage is increased if the liquid is agitated, and flows quickly over the pipe surfaces.

Suction-line controls

As in domestic refrigerators and food freezers, the suction lines of low temperature commercial and industrial plants incorporate suction line accumulators – now often termed 'liquid separators' – to prevent liquid refrigerant from flowing back to compressors. Their large cross-sectional areas cause any liquid in the line to expand and vaporise before it reaches the compressor suction port. The two additional refrigerant controls which follow, and a heat exchanger, may also be fitted in suction lines.

Evaporator pressure regulators

This control has much in common with a constant pressure expansion valve. Its inlet and outlet ports are on either side of a valve seat. Suction pressures must overcome the pressure exerted by a spring-loaded bellows before the seat can open and allow low pressure refrigerant to flow to the compressor.

It will be seen from the cross-section in

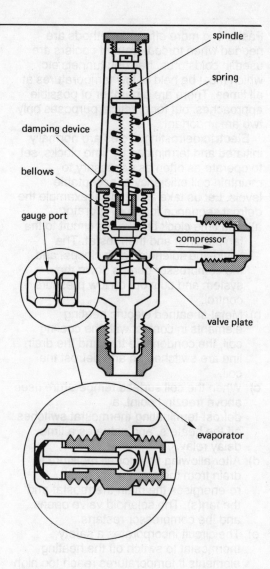

Fig. 6.31 Suction pressure control

Fig. 6.31 that if the evaporator pressure falls below the set point level, the port will be closed by the tensioning spring. If the evaporator pressure exceeds the set point, the extent to which the port will open – and the amount of gas which can pass through it – will vary with the difference between set point and evaporator pressures.

This type of control is used to prevent evaporator pressures (and therefore temperatures) from falling below a required level. It can prevent air cooling coils from frosting over, or water chillers from freezing up. When several evaporators are used with only one compressor or condensing unit, an evaporator pressure regulator in the suction line from each evaporator – before they enter a common suction header – will provide individual temperature control for each cooler. The compressor can then be cycled by a low pressure control.

Suction pressure regulators

The cross-section in Fig. 6.32 shows that this valve is similar to the pressure regulator described previously but that the refrigerant flow path is reversed and the valve seat opened or closed by a disc positioned on the other side of the port. The pressure exerted by the spring bellows is now used to open the port, unless the pressure at its outlet is above the set point value. If it is, pressure on the disc will overcome the pressure from the bellows, and the valve will be closed.

This type of valve will, therefore, prevent suction pressures from *exceeding* a pre-set level. It can be used to protect the compressor motor from excessive

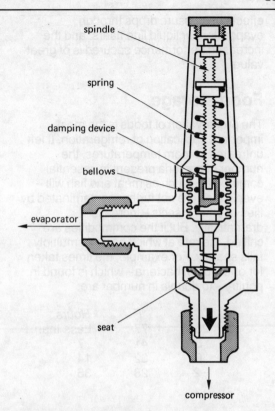

Fig. 6.32 Suction regulator

pressures which might be experienced when a system is first started, and has not pulled down to design temperature levels; or when an excessive load is imposed on the evaporator; or when a defrosting cycle has generated abnormally high pressures. It will not remain permanently closed, but will modulate and allow the compressor to reach a safe suction pressure in several steps without overloading the motor.

Modified suction-line controls

In large systems, pilot-operated pressure regulators can provide accurate control of suction-line conditions. A second spring is used to counteract the first, with both acting on a diaphragm, an internal pressure equaliser also being provided. This arrangement is similar to that of a C.P. expansion valve, as shown in Fig. 6.26, but the direction of refrigerant flow is reversed.

Evaporator pressure regulators are also available in designs which cause them to react to the effect of a temperature sensing bulb on a spring-loaded bellows. This regulates the position of a needle valve, which opens or closes a pilot orifice. The port divides high and low side system pressures; and the valve accurately controls the leaving temperatures of the liquid – or vapour – being chilled by the system.

Suction-line heat exchangers

A heat exchanger which reduces the temperature difference between liquid and suction lines can meet several needs.
a) To increase system efficiency when operating at low suction temperatures with R12 or R502.
b) To subcool liquid refrigerant and prevent the formation of flash gas.
c) To evaporate small amounts of liquid refrigerant which might have entered the suction line (this being the only reason for which heat exchangers are normally used in R22 systems).

One of several designs can be used:
a) Liquid and suction lines soldered together (common practice in domestic

refrigerators) with the liquid line at the bottom. This is only done on small systems.
b) Shell-and-coil arrangement, with liquid refrigerant in the coil, which can be finned to increase heat exchange capacity. When using this arrangement it is desirable for metering and solenoid valves to be added, to enable any oil accumulating in the shell to be returned to the compressor crankcase.
c) Concentric tube-in-tube layout. This is not as efficient as shell and coil, but avoids oil return problems.
d) Spiral wound interchangers, with the liquid line wound around a finned length of suction line. This type is most often used with forced-draught evaporators used in low temperature coldrooms.

Although refrigerant heat exchangers have been described under the general headings of suction line fittings, their benefits when used in refrigeration systems are not confined to the low side. The liquid sub-cooling does much to reduce the effects of pressure drops through evaporators or liquid line risers, and the increased performance secured is of great value.

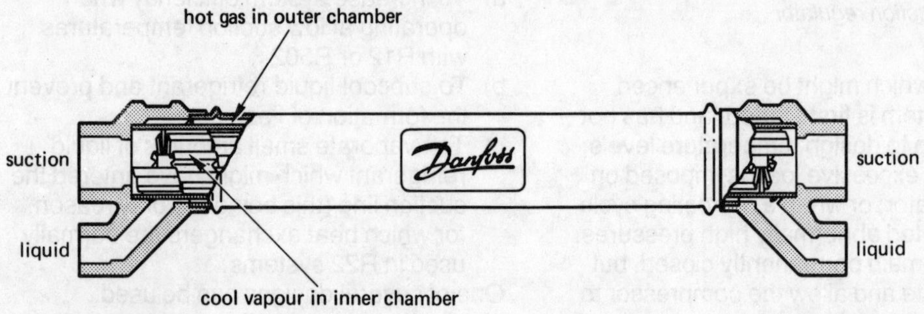

Food storage

The preservation of foods is the most important application of refrigeration. If left untreated at room temperatures, the number of bacteria present in essential commodities such as meat and fish will – even if they are not further contaminated by flies or other insects – increase dramatically. But if the commodities are chilled, the rate at which bacteria multiply falls sharply. For example, the times taken for one type of bacteria – which is found in poultry – to double in number are:

°C	°F	Hours
25	77	Less than 1
5	41	7
0	32	14
−2	28	36

Progressive deterioration in the condition of foodstuffs results in damage which can be seen, or smelt. For example, bacterial growth on the surface of beef or chicken joints eventually causes the formation of slime if the meat is moist and mould if it is dry. To multiply, bacteria must have access to oxygen, and oxygen itself can cause damage not related to bacterial growth. A common example is that fish which has a high oil content will, when exposed too long to air, smell rancid as the result of oxidation of the oils; non-oily fish does not. However, long before bacterial action has reached the point where it causes foods to change colour, be coated with moulds or slime, or smell rotten, the really dangerous bacteria – such as salmonella – will be numerous enough to poison, and perhaps to kill, anyone unfortunate enough to eat the food on which they are present.

It is obviously desirable that perishable foods be chilled as soon as they are prepared, and stored at the coolest possible temperature. Unfortunately we cannot solve the problem by deep-freezing them all. To do so would kill many bacteria, but by no means all; and the survivors would again begin to multiply if and when temperatures were allowed to rise. This is why it is dangerous to allow frozen foods to thaw and then to refreeze them. Every time this happens, bacterial content increases and can have very serious consequences when the foods are finally eaten. Even heating foods to temperatures at which most bacteria will certainly be killed – as when milk is pasteurised for instance – does not provide a 100 per cent guarantee of safety after it has returned to a more normal

Fig. 6.33 Spiral-type heat exchanger

temperature level. Surviving bacteria will start to grow once more, and there is the ever present danger of further contamination when foods are handled, breathed on or cut or minced. Even pasteurised milk should be stored at a temperature close to its freezing point.

There is no single 'best temperature', or set of ideal conditions, at which all foods should be stored; what suits one may be damaging to another. When foods are held at temperatures above freezing point, humidity is important: some foods are quickly dehydrated and become inedible if held in too dry an atmosphere, whilst others deteriorate if the humidity is too high – as we have already seen, bacteria flourish on moist surfaces. The velocity at which air is circulated over unwrapped, chilled foods is another critical consideration. If it is too high the surfaces of cut meats or fish may dry out and look unappetising, and weight lost by dehydration is money lost by traders selling produce by the pound or kilo.

Some examples of recommended storage conditions and average physical properties of a number of common foods are detailed in Fig. 6.34. Note that many have a high water content – and therefore much latent heat must be to be removed if they are frozen. The freezing process consists of three steps:
a) chill to freezing point,
b) remove latent heat,
c) sub-cool to storage temperature.

There is some overlap between stages, because latent heat does not emerge in a neat block, but total heat to be removed is the same. Many fruits and vegetables, which are living organisms, continue to give off heat even after being chilled (but not frozen) and placed in a coldstore. This heat – which is greater at higher storage temperatures – must be allowed for when heat loads are calculated and refrigeration equipment selected. It is the result of fruit, in particular, ripening; and in some cases

Product	Storage Temp. °C	Humidity RH%	Freezing point °C	Specific Heat kCal/kg/°C Unfrozen	Frozen	Latent Heat kCal/kg	Respiration kCal/kg/ 24 hr.
Apples	1	90	−1	0.86	0.45	67	0.40
Bananas	18	90	−1	0.80	0.42	60	2.33
Beef	1	90	−2	0.77	0.40	55	
Butter	4	80	−1	0.64	0.34	8	
Cabbage	0	92	−1	0.94	0.47	73	0.80
Carrots	0	95	−1.5	0.86	0.45	70	0.96
Cheese	0	65	−13	0.64	0.36	49	1.30
Dates (cured)	0	70	−16	0.36	0.26	16	
Eggs	−1	80	−2	0.76	0.41	56	
Fish							
(fresh, iced)	0	95	−2	0.76	0.41	56	
(dry)	2			0.56	0.34	36	
Ice cream	−26		−5.5	0.78	0.45	53	
Lamb	1	85	−2	0.67	0.30	46	
Lard	7	90		0.52	0.31		
Lettuce	1	95	−0.2	0.96	0.48	76	2.05
Margarine	2	65		0.48	0.34	19	
Melons	5	85	−0.5	0.94	0.48	73	0.97
Milk	0		−0.5	0.93	0.49	69	
Onions	0	70	0.8	0.91	0.46	69	0.28
Pineapple	7	85	−1	0.88	0.45	68	
Peppers							
(fresh)	7	90	−0.7	0.94	0.47	73	0.42
Potatoes	5	90	−0.7	0.82	0.43	62	0.43
Poultry	0	85	−3	0.79	0.37	59	
Shrimp	0	95	−2	0.81	0.43	61	
Tomatoes	7	85	−0.5	0.95	0.48	74	0.45
Yams	16	85		0.79	0.42	59	

Fig. 6.34 Storage conditions for common foods (not frozen)

this process is delayed by storing them in an atmosphere of *inert gas*, so that there is no oxygen present to encourage chemical reactions necessary for ripening.

Ideally, recommended storage temperatures and humidities should be closely adhered to – the preferred tolerances would not exceed plus or minus 1·1 °C (2 °F) or 3 per cent RH. In practice the majority of coldrooms contain several different types of food and compromises must be made. Most errors are made in the direction of too high a temperature – frozen foods in particular should not be allowed to rise above – 18 °C (0 °F) – and too low a relative humidity.

The principal factors governing relative humidity in a coldstore are the temperature of the evaporator and the vapour pressure of the air (which can change as the result of moisture being yielded by the food, workmen or infiltrating air). As can be confirmed by a psychrometric chart, the lower the evaporator temperature the more moisture it will remove from air passing over it. It is generally accepted practice to aim for a coil temperature approximately 5·5 °C (10 °F) below room DB in stores operating down to freezing point, and 8·4 °C (15 °F) below room DB in rooms holding frozen produce. Once it has been frozen – and often packaged – food cannot have its moisture content reduced. For foods held above freezing point, several equipment manufacturers produce 'high humidity' evaporators. These have larger than usual cooling surfaces and circulate smaller than average air volumes, but must still be used at proper temperature differences.

Having looked at the main equipment components we shall use and the storage conditions which they should maintain, we can now review some of the applications and the types of cabinet which we can expect to install and service. The items included, and many of the comments made, are based on practices followed in warm and humid, or very hot and dry climates. These impose stresses, and require approaches which are often unnecessary in more temperate regions. They will be pointed out if they are likely to present problems to an engineer during the course of a typical day's work.

Refrigerated display cases

A very wide range of cabinets of different shapes and sizes are available and all have one thing in common. To operate efficiently and hygienically in tropical regions, they need to be installed in air-conditioned stores! The basic summer design condition used by American manufacturers is 24 °C (75 °F) at 55 per cent RH; and in the interests of manufacturing and operating costs, some designers use external conditions even less likely to require the provision of anti-condensate heaters. If possible, therefore, always obtain manufacturers' performance ratings; they may explain unexpected symptoms and help establish such factors as the frequency and length of defrost cycles.

We have space to describe only fully enclosed and fully open cabinets. These may have self-contained refrigeration systems or (more usually) may be arranged for use with remote condensing units. At this stage we will consider only the cabinets, and leave condensing unit installation needs for a later chapter. In view of earlier comments on bacterial contamination of foodstuffs, we must ensure that all display equipment is maintained at peak efficiency. This especially applies to open cabinets which can be affected by the heat radiated from lighting fittings, draughts created by fans or inconveniently placed air conditioning grilles, and possibly also by incorrect loading.

Meat display cases

Meat and meat products which are not pre-packaged for self-service are invariably displayed in fully enclosed cabinets designed for rear loading and service by store personnel. A typical cabinet cross-section is illustrated in Fig. 6.35. Note that this is one of the few cases in which a high-mounted finned evaporator cools the cabinet and its contents by gravity circulation of air, i.e. convection currents.

Packaged meats in self-service areas will be displayed in open cabinets, which can be either low level 'well' or 'island' types, or of multi-shelf 'wall' design. Cross-sections illustrating the basic features of both types are shown in Fig. 6.36(a) and (b), and it will be seen that forced-draught air circulation is used. This provides efficient air movement through evaporators and over the produce, and also produces an 'air curtain' between the meat and the air outside the cabinet.

It is vitally important that contents are

COMMERCIAL AND INDUSTRIAL REFRIGERATION

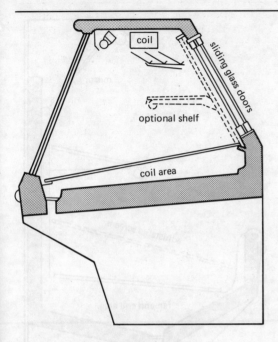

Fig. 6.35 Closed meat display case

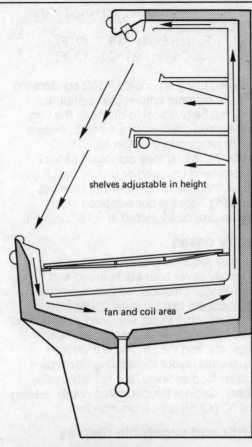

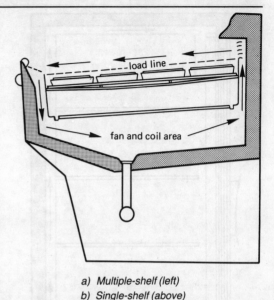

a) Multiple-shelf (left)
b) Single-shelf (above)

Fig. 6.36 Open meat display cases

kept below load lines and that nothing is done to obstruct the air streams. The surface temperature of the meat must not rise above 4·4 °C (42 °F) if it is to have a good saleable life and retain a good colour.

Cabinet designs usually have foamed polyurethane insulation between stainless steel, aluminium, or plastic inner liners and stove-enamelled or stainless-steel outer skins. Robust construction, and some form of inbuilt 'bumpers' to protect cabinets from carelessly handled food trolleys are essential. Cabinet sections are normally available with only one pair of ends to complete any number of units, providing uninterrupted display. The space beneath display trays in both rear-service closed-type and well-type self-service models can house condensing units. Reference to the external design conditions quoted earlier may indicate the necessity of either emptying cabinets, or using night covers to protect their contents, if air-conditioning systems are switched off overnight. Positive defrost arrangements – generally electric – are essential, and condensate must be properly plumbed (using suitable traps) to a drain.

Low temperature displays

Cabinets to display self-service frozen foods or ice cream are most often of the 'well' type illustrated in Fig. 6.36(b), or of upright 'wall' design incorporating fully glazed doors to ensure that packages are always clearly visible (see Fig. 6.37).

Doors invariably have triple-glazed panels with all air removed from the spaces between glass panes. Forced-draught evaporators discharge air which forms a 'curtain' preventing humid air from freezing design air temperatures approved in the USA are:

    Frozen foods −18 °C (0 °F)
    Ice Cream −25 °C (−13 °F)

When using air-cooled R502 condensing units in extreme ambient temperatures, it may not be practical to maintain the very low suction pressures required to secure an air temperature as low as −25 °C (−13 °F). Whilst they obviously cannot recommend unsound or unsafe practices, the authors have not heard of illnesses resulting from the consumption of ice cream previously stored at −18 °C (0 °F).

Dairy cases

Display requirements for daily produce normally cover both staff-served and self-service products, so that two types of cabinets are used in most major stores. The first is the totally enclosed, rear access type, generally as described for meat cabinets; and the second is a vertical multi-shelf layout resembling wall-type frozen-food cabinets, but not fitted with doors. Cabinet temperatures not exceeding 4·4 °C (42 °F) are recommended.

Fruit and vegetable display

For these products, self-service is encouraged and wall-type, multi-shelf cabinets are normally used. Since not all types of produce need to be refrigerated, cabinet designs often compromise between well- and wall-type cabinet features, with the bottom display area held at 7·2 °C (45 °F) and the upper shelves not refrigerated. A typical cabinet cross-section appears in Fig. 6.38.

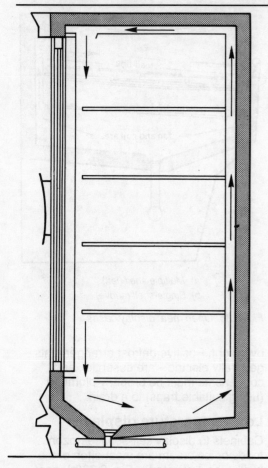

Fig. 6.37 Glass door display case

on the cabinet contents when doors are opened. Door gaskets have electric heating wires to prevent them freezing shut. Automatic electric or hot gas defrost is essential, with attention paid to air ducts and fans as well as evaporators. Internal

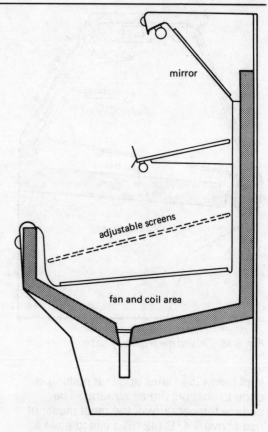

Fig. 6.38 Fruit and vegetable display

Fresh fish display cases

Fresh fish should be maintained as close as possible to freezing point, in a humid atmosphere not affected by high air velocities which would dry out the produce on display. Many merchants favour having the fish arranged on trays of flake or crushed ice, or on refrigerated slabs.

Compromise designs are once again common. Cabinets are usually arranged for back- or top-service access, and have finned cooling coils mounted as high as possible on the rear wall to cool the contents by air convection currents. In some cases the bottom of a stainless-steel or marble display slab is refrigerated by pipe coils attached to its underside. In others, the slab is constructed from roll-bonded aluminium made in similar fashion to the evaporators of domestic refrigerators. Unless storage compartments are used underneath the display area, the slabs are well insulated beneath this evaporator and contain properly trapped drainage systems. Internal design temperatures are usually 1 °C (34 °F).

This subject has been covered at length because display applications are common, and because methods illustrate many of the basic rules used in other forms of short-term storage cabinets – even coldrooms, or 'walk-ins', which we will look at next.

Coldrooms and cold storage

Until recently coldstores were, regardless of their size, 'built-in'. They were constructed by applying slabs of expanded polystyrene, cork or other insulating material to a wall which was first vapour-sealed with, usually, bitumastic materials. Insulation was arranged in several thicknesses, with joints staggered to reduce direct through-paths for heat, and set in hot- or cold-pour bitumen or other adhesives. Where necessary, it was supported by skewers attached to the outer wall, or by timbers dividing each surface into reasonable areas. After further vapour-sealing, the interior was finished. Popular approaches included covering the insulation with asbestos cement or similar sheets, secured to timber supports and with all joints covered; or lining the insulation with chicken wire secured to timber uprights and using this as the basis of a rendered and tiled inner surface. Ceilings were treated in similar fashion. Floors were laid in *styrene*, or other insulating material able to carry heavy loads, and covered with two inches or more of Portland cement. This was coved to the walls to prevent the entry of water.

The majority of modern coldstores are assembled from factory-made, standard-dimension panels, the best having quick-make jointing mechanisms which produce air- and vapour-proof joints between panels. They are available in expanded polyurethane up to 125 mm (5 ins) thick. This is foamed between metallic skins of galvanised or stainless steel, or aluminium, or plasticised sheet metal. The top skins of floor sections are thick enough for normal use, but where heavy loads on a small area result from the use of forklift trucks or other mechanical handling equipment the use of concrete wearing floors is necessary. Doors are supplied ready-hung in wall sections, several alternative designs being offered.

A typical small walk-in room of this type is illustrated in Fig. 6.39. This shows the relationships of floor, wall, and ceiling panels, and the positions in which they are

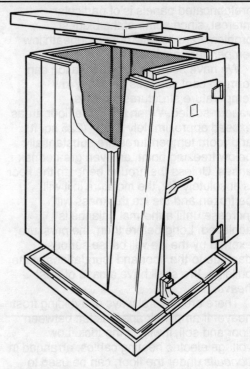

Fig. 6.39 Sectional walk-in

locked together. This type of construction can be used for large as well as small buildings, subject to heights not exceeding approximately 7·5 m (25 ft). When the rooms are more than say 4·4 m (14·6 ft) wide, or wall panels are more than one tier high, it is necessary to support the insulated panels with structural steelwork. Heavy equipment (such as carcass rails) need to be independently supported, and not hung from the insulation.

A cut away diagram (Fig. 6.40) illustrating a refrigerated building built from prefabricated panels is of particular interest, since some of the biggest problems are literally buried beneath low temperature coldstores.

We have seen how condensation can form on cool surfaces, and why low temperature structures must be vapour-sealed. When coldstore floor areas exceed approximately 21 m² (225 sq. ft) and room temperatures are substantially below freezing point, an even greater risk arises. Unless the ground beneath the floor is absolutely dry, the moisture in it will be frozen and the ice thickness will increase until a thermal balance is achieved. Long before then, the pressure exerted by the ice will cause serious damage to the floor and foundations of the building. We shall have a case of *frost heave*.

There are several ways of avoiding frost heave. If ample air space is left between floor and soil, it will do the trick. Low voltage electric heating cables, arranged in *conduits* under the floor, can be used to prevent the ground temperature from being reduced below freezing point. A more simple precaution is the use of 150 mm (6 in.) pipes at approximately 1·8 m (6 ft) intervals through which air is circulated to keep the subsoil above freezing point. These pipes should be installed between floors and subsoil, as indicated in Fig. 6.41. If perforated, they can be used for drainage purposes and arranged to discharge into sumps. If pipe length exceeds 9 m (30 ft) mechanical air circulation is desirable, warm air vented from condensers or plantrooms being a source of cheap heat where air temperatures may fall beneath desirable levels. Mesh screens must be used to keep animals etc. from entering the pipes.

Necessary accessory items include internal handles to all walk-in doors to prevent personnel being accidentally locked inside. Stable-type doors (i.e. ones that are split horizontally so that only one half need be opened) or serving hatches help minimise infiltration of ambient air. So do air curtains mounted outside and above doors, or the use of self-closing synthetic rubber or plastic *crash doors* or curtains mounted just inside the main entrance door. Pressure relief ports prevent excessive air pressures resulting if rooms

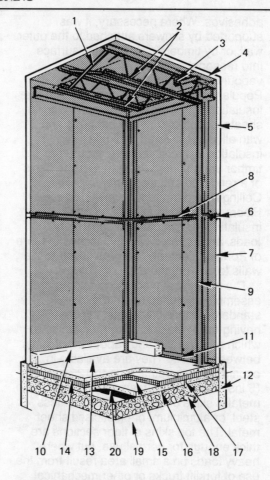

1 Lateral bracing as required
2 Web joists in required size
3 Ceiling anchors, spaced every 92" on every supporting joist
4 Prefabricated ceiling panels
5 Prefabricated upper vertical panels
6 Joining adaptors to join two vertical panels in height
7 Prefabricated lower vertical panels
8 Horizontal steel struts
9 Load-bearing perimeter steel columns
10 8" high × 6" deep reinforced concrete curb, poured after erection of building
11 Floor screeds—beneath all vertical panels when insulation is built-in
12 Footings extend below frost line
13 4" thick reinforced concrete wearing floor
14 2 layers of 2" thick built-in sheet urethane. Staggered joints
15 4" thick reinforced concrete sub-slab
16 2" rock aggregate, 15" deep
17 1" drain holes for water seepage
18 4" or 6" dia. perforated drain and ventilation pipes spaced on 6 ft. centres, running the entire length of refrigerated building floor
19 Hard tamped earth
20 Watertight seal

Courtesy of Bally

Fig. 6.40 Coldstore erection detail

have to be brought up to temperatures well above long-term operating levels. If desired, door-operated electric microswitches can be used to control operation of lights, or to cycle the evaporator fans in small stores. It is however more satisfactory to arrange the evaporator so that cold air is not blown out of opened doors. *Dunnage battens* of timber, or steel bars, can be arranged around walls to ensure that air can circulate freely around foodstuffs, and to prevent damage to walls by mechanical handling equipment.

In addition to prefabricated insulated panels, the use of factory-assembled and charged refrigeration systems is increasing. A typical example of such a unit – they are available in sizes up to 5·5 kW ($7\frac{1}{2}$ hp) at least – is shown in Fig. 6.42.

The high side of the system is in a weatherproof housing, with the evaporator mounted on brackets which extend within the coldroom; brackets and refrigerant lines enter the room through notches cut at the top of wall panels. The ease with which speed-locked panels can be released and removed enables self-contained units to be removed or replaced without difficulty, so long as room heights are reasonable. Unit components differ little from those of a system normally assembled on site, but factory assembly enables installation methods to more closely resemble those used on room air-conditioners.

Refrigeration cooling loads

The cooling load on a system serving a walk-in coldstore is made up by a number

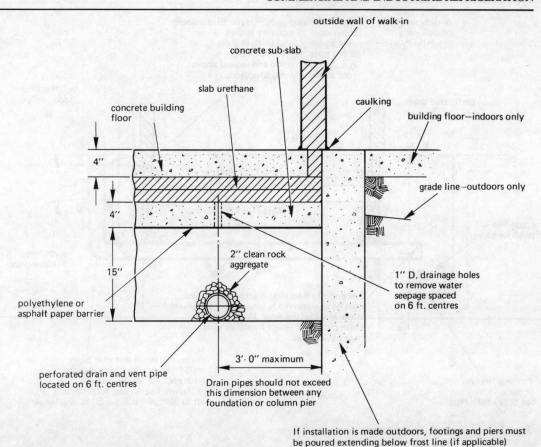

Fig. 6.41 Coldstore floor detail a) Section through outside wall

of factors, not all of which apply to other classes of equipment. They are as follows.
a) Heat gains through walls, floor and ceiling.
b) Heat required to cool and dehumidify air entering when doors are opened.
c) Heat given off by lights, evaporator fan motors, and any forklifts or other plant used in the store.
d) Latent heat and sensible heat emitted by staff working inside the store.
e) Heat to be removed to cool entering

TROPICAL REFRIGERATION AND AIR-CONDITIONING

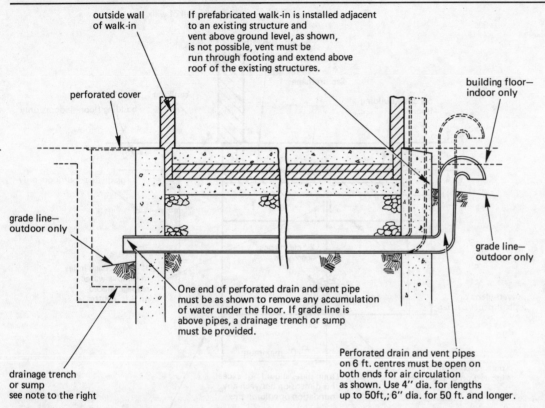

Fig. 6.41 b) Section through length

Fig. 6.42 Self-contained refrigeration system

foodstuffs down to storage temperature; or, in the case of freezers, to chill, freeze and subcool the produce. Whilst installation or service engineers are seldom asked to make detailed calculations of heat loads, you need to know how they are built up, and be able to remedy any cause of abnormally high loads. You will in practice need to be able to advise users on loading procedures, to permit free air circulation around room contents and avoid loading too much or too warm produce in so short a period of time that the refrigeration equipment is heavily overloaded. You may find it helpful to study local health regulations, which might preclude the use of structures which do not have inner liners which are free from jointing cracks. Finally, you must know that each room must have proper drainage facilities, including trapped (and where necessary heated) drainage lines; so that it can be washed or hosed clean of dirt, blood, or other contaminants.

Block ice plant

The applications described earlier in this chapter will all have required the use of R12 or R502 for, respectively, higher and lower temperature levels. We should now look at an essential application which usually operates on R717: block ice manufacture.

Ice production normally continues day and night, cans of fresh water being frozen in a tank of brine. The steel tank is heavily insulated, and the brine cooled by a submerged evaporator at one side, or one end, of the row of cans. An agitator is fitted at one end of a raceway, to circulate brine over the coil and between the cans. This increases the heat transfer capacity of all the metal surfaces. The cooling coils are of the flooded type, with the supply of refrigerant metered by a float valve. The suction line has an accumulator, or separator, to prevent liquid refrigerant being drawn into the compressor. This is of open design, driven directly or through vee belts, from an electric motor. In some cases diesel or steam engines are used as prime movers – usually where power supplies are limited. Most condensers are of the water-cooled, shell-and-tube type, operating with cooling towers and circulating pumps. Evaporative or air-cooled condensers are, however, also available.

A typical ice factory layout is illustrated in both plan and elevation in Fig. 6.43. It will be seen that cans are lifted in groups, and moved by an overhead travelling crane. Those removed full of ice are dipped into a thawing tank, enabling ice blocks to slide from the cans when they are subsequently tipped onto their sides. The ice tip is close to a store, in which several days' ice production can be held at approximately $-4.4\ °C\ (24\ °F)$. The ice-store evaporator is normally of the forced-draught type, operating either on R717 or on brine pumped from and returned to the main tank.

Ice cans are emptied, then refilled and returned to the tank, during a normal eight-hour working day. The mass of the brine evens out load fluctuations, and usually enables compressor and motor to

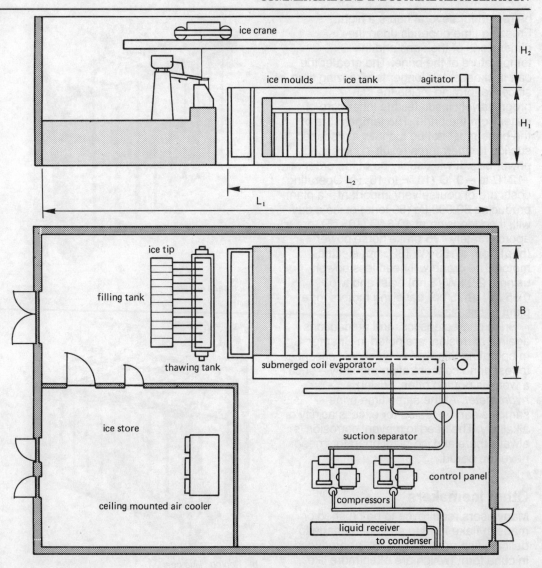

Fig. 6.43 Block ice plant

be rested for several hours a night. Freezing time depends upon the dimensions of the cans and the temperature of the brine. The greater the can thickness, the longer the freezing time since ice forming inside the can progressively insulates the inner core of water from the brine. The temperature of the brine must not be too low, as reduced suction temperatures result in increased power bills. It will usually be of the order of −12 °C to −9 °C (10 °F to 15 °F). Operating costs are of course very important – a plant producing 60 000 kg (50 tons) of ice a day will, if condensing at 40·6 °C (105 °F), draw approximately 175 brake horse power (bhp) from a 168 kW (225 hp) electric motor. The crane will need less power, using a 2·2 kW (3 hp) hoist and two 0·55 kW (0·75 hp) travelling motors for a plant of this size.

Brine characteristics, and precautions against corrosion, are noted in Chapter 4. In operating this type of plant, it is essential to test brine concentration regularly – once a week is not too often – using a *hydrometer*. At the same time brine samples can be tested for excess acidity or alkalinity. The need to prevent corrosion is always there and ice-plant operators need never be bored.

Other icemakers

Many users require ice to be crushed or made in flake form to prevent damage to delicate fish or other tissues, or produced in cube form (which are often more like cylinders!) for use in drinks. It is not proposed to examine structural details of these different designs in depth, since so many are in common use and all are very complex. Most, however, use R12 and freeze water in a refrigerated tray before the cubes are cut by electrically heated wires. Operating cycles are usually controlled by time clocks, and liable to go wrong if power voltage and/or periodicity values are not kept steady. Problems can result if scale forms on surfaces exposed to water, and the use of special descaling chemicals is then necessary. Most machines have built-in ice-storage bins to receive cubes at the end of each freezing cycle. The installation and maintenance instructions must be closely followed.

Food freezing

High quality frozen products cannot be processed on an industrial scale using normal coldstores operating at or about −18 °C (0 °F). The plant used must be capable of much lower temperatures, to freeze the produce as rapidly as possible. Slow freezing often causes the formation of large ice crystals, which rupture tissue cells and result in the loss of both flavour and quality. A vegetable with a high water content that is frozen slowly will be very soggy when it is warmed up again. We must think in terms of temperatures as close as possible to minus 40 °C (−40°F as well!) to ensure excellence.

Food can be frozen in several different ways: by using blast freezers, either static form or in 'tunnel' form (in which food moves along on conveyor belts); by horizontal or vertical plate freezers; by immersing fish in low temperature brine; by

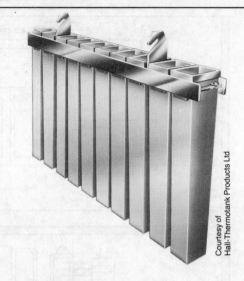

a) *Ice cans in frame*

b) *Tipping/filling gear*

Fig. 6.44 *Ice making equipment*

spraying foods with, or immersing them in, halogen family refrigerants – which are later recovered, and recondensed; by exposing them to liquid nitrogen or liquid carbon dioxide at temperatures of −60 °C (−94 °F) or lower. The latter methods are not much used in our part of the world, and we can confine our present study to blast freezers and plate types.

Figure 6.45 is a cutaway sketch of a blast freezer. At one side of the insulated structure, a battery of fans forces large volumes of air through very low temperature cooling coils, and then over (in this case) pre-packed foods. These are arranged so that as many external surfaces as possible are directly exposed to the blast of extremely cold air. We said earlier that in high ambient temperatures it is difficult to achieve temperatures as low as minus 40 °C at the evaporator, but this time we must try hard to do so!

If racked packages remain still, we have a blast freezer; if they are moved, manually or mechanically, through the freezing area, the equipment can be described as a tunnel. Foods need not, of course, be packaged before being frozen. This is a matter of convenience, both when freezing and when the product is marketed. If the sizes of food portions are regular – as with peas – it is convenient to freeze them before they are packed into plastic bags. This also helps prevent the formation of lumps of produce. If the sizes are not regular, or the food is normally sold in standard size cartons, plate freezers are more often used. Whilst being frozen, cartons are then held firmly between the plates, preventing distortion and providing a neatly shaped package.

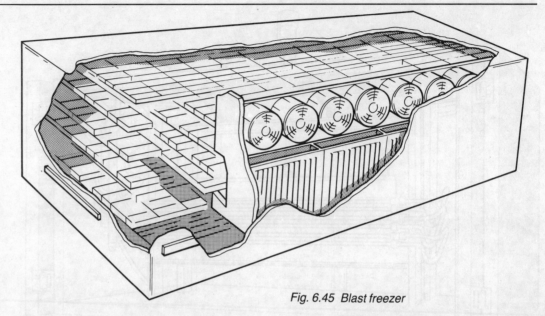

Fig. 6.45 Blast freezer

Figure 6.46 illustrates a typical horizontal plate freezer. (Vertical plate models are used mainly to freeze fish aboard trawlers.) Note that the cabinet is of the pass-through type, and is well instrumented. (We shall have more to say about instrumentation standards in later chapters.) The plates can be moved apart – refrigerant connections being made through flexible lines – for loading and then closed to a precise spacing by an hydraulic mechanism. The final distance between plates is about 50 mm (2 ins), and with the plate surfaces extremely cold, very fast freezing results. Times do of course vary with the nature of the food, its thickness, and the operating temperatures.

Low temperature refrigeration systems

Before leaving the subject of low temperature work, we must understand the methods used to span the gap between normal condensing temperatures and extremely low suction pressures. Apart from the use of liquid refrigerants, liquid nitrogen, etc. (technically termed *cryogenic* processes) two procedures based on adaptations of standard compressors are widely used.

Two-stage (compound) compressors

Open, reciprocating compressors are redesigned so that after having been compressed in two or more cylinders, high pressure gas leaving the discharge port is cooled by injection of liquid refrigerant or passage through a heat exchanger, and then enters a further, high pressure cylinder in the same crankcase. It is here compressed even more, before being discharged at design head pressure.

The very high compression ratio results in much heat being generated, and this type of compressor is fitted with water-cooled jackets above and around the cylinder heads if operating on R717.

R12, or R22 can also be used in this type of compressor, which is made only by specialist manufacturers. In view of the extreme operating pressures, particular attention must be paid to lubrication arrangements as well as the construction of suction and discharge valves, and any unloading devices used to secure capacity control.

Injection cooling valves are also specialised, consisting – in over-simplified terms – of a strainer through which liquid refrigerant passes into a pressure-operated two-way throttling valve, connected in turn to the inlet port of a TEV. The operation of the TEV is governed by pressures sensed by a bulb mounted between the low pressure stage discharge manifold, and the high pressure stage suction port.

Incidentally, expansion valves used in low temperature systems are not radically different from those we have already examined, unless they are required to operate below about −50 °C (−60 °F). TEVs rated down to −100 °C (−150 °F) still operate on the same principles, but are precision built.

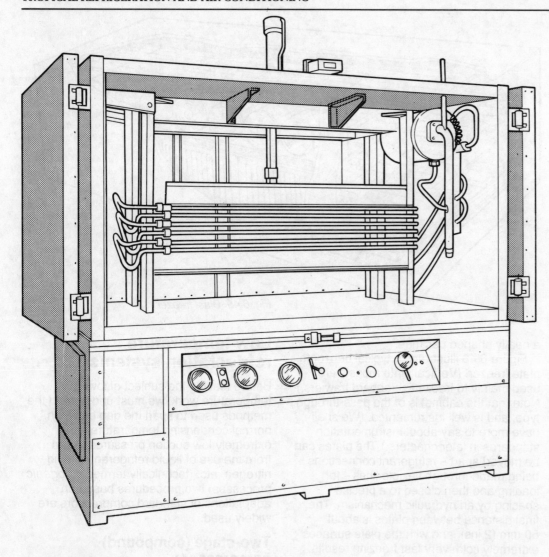

Courtesy of Jackstone Froster

Fig. 6.46 Plate freezer

COMMERCIAL AND INDUSTRIAL REFRIGERATION

Cascade systems

Whereas compound compressors are designed to compress gas in two separate steps within a single crankcase, the cascade system uses two separate refrigeration systems. This is costly, and not normally required for other than laboratory or specialised industrial processes demanding extremely low temperatures.

The principles of the cascade system are illustrated in Fig. 6.47. The evaporator of one system is one half of a heat exchanger, of which the other half acts as the condenser of the second, lower temperature circuit. A wide gap between extreme suction and condensing pressures can thus be bridged in two steps, greatly reducing the compression capacity required of each compressor. The lower temperature circuit in particular has a much easier job, since its condensing pressure is about the same as the suction pressure of the higher temperature system.

A second benefit can be gained – there is no need to use the same refrigerant in each circuit. One suitable for medium pressure – R12, for example – could be used in the higher temperature system, and one of the low temperature refrigerants in the low temperature circuit.

Note one source of possible confusion – the practice of calling the lower temperature of the two systems the 'first stage'. It would not be illogical to think of it as the second! We suggest that readers think of, and describe cascade systems in terms of higher and lower temperature circuits.

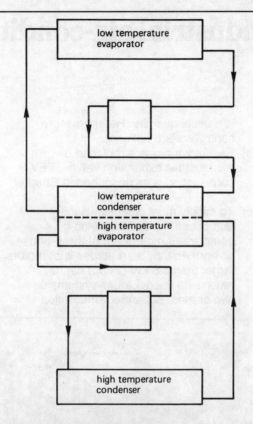

Fig. 6.47 Cascade refrigeration

Refrigerant pumps

One further brief comment on large, direct-expansion installations, in which very long pipe runs or very high risers might cause excessive refrigerant pressure drops. Booster pumps can be installed in liquid lines to maintain adequate pressures. This is most often done in large R717 direct-expansion systems, or where pressure drops in risers make it difficult to maintain steady R717 levels in flooded evaporators. We'll have more to say about pumps in the next chapter – but in the context of water, not R717. This is not of great importance – so long as you know why refrigerant pumps are used, and in what type of circuit, we need not worry about the finer details of their construction.

Revision questions

1. a) Sketch the construction of a pressure-operated water valve for the shell-and-tube condenser of a 3 kW R22 system.
   b) Described in detail the method of setting the valve in (a) above to ensure maximum economy in both power consumption and the use of water.
   c) If the evaporating temperature is −5 °C, give typical values for condensing pressure and condensing water outlet temperature if the inlet water temperature is 24 °C.
   (WAEC)

2. a) Make simple line sketches showing vertical cross-sections through, and label the main components of:
      (i) a water cooling tower,
      (ii) an evaporative condenser.
   b) Explain the operating principles of each of the above.

7. Commercial and industrial air-conditioning equipment

As with refrigeration equipment, studies of commercial and industrial air-conditioning plant can be based on practices used in domestic appliances, with which we are already familiar.

Packaged comfort-cooling units

Apart from changes in size few variations need to be made to the domestic types described in Chapter 5 to suit them for commerial applications. For self-contained units with nominal capacities between 17·5 and 53 kJ/s (5 and 15 TR) the principal changes are as follows.

a) The use of separate evaporator and condenser fan motors. Most designs are based on evaporator air volumes of 55 ltr/s per kJ/s (400 cfm/TR); but individual needs can be perhaps 15 per cent smaller or greater, and fan speed control is necessary. It normally takes the form of a variable-pitch motor pulley, forming part of a vee belt drive kit. For their part, condenser fans will probably be arranged for vertical discharge, to limit fan noise, or, for the same reason, be replaced by centrifugal types. Condenser air volumes still vary with heat exchanger surface areas and design condensing temperatures; but 110 ltr/s per kJ/s (800 cfm per TR) is a usual quantity. Head pressure controllers are optional.

b) Capillary tubes are replaced by thermostatic expansion valves. TEV construction is as described in Chapter 6.

c) To satisfy most electricity supply authorities, welded hermetic compressor motors are of three-phase design. As they start across line, motors larger than 5·5 kW (7½ hp) cannot always be used. Larger systems use two or more separate refrigeration circuits, with time delay relays to prevent more than one compressor motor from starting at once.

d) The use of remote, room-mounted thermostats may be necessary. For safety reasons, these should operate on 24 V supplies, control voltage transformers being fitted inside the cabinets. Care must be taken to locate the sensors of the thermostats where they will not be screened by curtains etc., or affected by external heat sources.

Fig. 7.1 Horizontal self-contained packaged unit

COMMERCIAL AND INDUSTRIAL AIR-CONDITIONING EQUIPMENT

e) Good practice requires the inclusion of liquid-line solenoid valves in pump-down circuits.

A typical horizontal self-contained packaged unit for outdoor use is shown in Fig. 7.1. As with domestic product ranges, split systems are available (see Fig. 7.2), outdoor condensing units operating with either vertical or horizontal evaporators.

All manufacturers follow broadly similar design trends, and to minimise installation complications, many offer precharged and pre-insulated liquid and suction lines which terminate in self-sealing, quick-connect joints. Electric or hot water heating options are always available; and in many cases heat-pump versions are also offered.

A few manufacturers also offer completely self-contained, vertical air-conditioners for use in indoor plantrooms. Their air-cooled condensers require the use of more complex ductwork than conventional designs and it is always difficult to handle large volumes of air within confined spaces, without making a lot of noise. These factors appear to have inhibited sales of what would otherwise be an extremely attractive product.

In addition to air-cooled equipment, very similar product ranges with water-cooled condensers are available from most manufacturers. Depending on price and quality, the condensers are of tube-in-tube, shell-and-coil, or shell-and-tube design. Condenser details are as described in Chapter 6.

Considerably larger packaged air-conditioners are of course available; capacities normally extend to 175 or 210 kJ/s (50 or 60 TR) or above. These normally use two or more accessible hermetic compressors, in separate circuits. Compressor motors have reduced voltage starting arrangements and the use of multi-step thermostats gives good capacity control potential.

a) Condensing unit

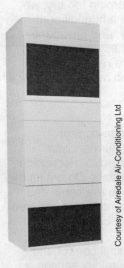

b) Airhandling unit

Fig. 7.2 Split system

Performance characteristics

Comfort-cooling models are designed to produce room conditions acceptable to the great majority of users. Acceptable conditions are not simply comfortable ambient temperatures but are a function of a combination of air DB, relative humidity, and also air movement. These can be plotted on charts, and form the basis of *effective temperatures* (ETs). At any given DB, the ET will fall if RH is reduced. The same result will follow a decrease in DB temperature, or an increase in the velocity with which air is moved. The great majority of packaged equipment is designed for, and performance rated at, internal conditions of 26·7 °C DB, 19·4 °C WB, 50 per cent RH; which produce an ET of 23·3 °C (80 °F DB, 67 °F WB, 50 per cent RH = 74 °F ET). These internal conditions are related to ambient temperatures – which have a direct bearing on suction and discharge pressures – of 35 °C DB, 23·9 °C WB (95 °F DB, 75 °F WB). This is much cooler than experienced in tropical deserts, and much less humid than at sea level in tropical rain forest regions. Although comfort-cooling equipment does not contain facilities to regulate internal humidity, good selection practices produce results which are acceptable in either type of climate. They approximate 24·4 °C DB, 18·3 °C WB, 21·7 °C ET (76 °F DB, 65 °F

WB, 55 per cent RH, and 71 °F ET). Unless otherwise instructed, this is what you should aim at when balancing an installation.

Close control air-conditioning

In a number of instances it is essential that relative humidity is closely controlled. This is the case with computer suites, the manufacture of medicines, the storage of steel 'belt' material to be used in radial tyres and a range of other applications. In some cases, too high a humidity would ruin an end product; in others, too low a RH can be very dangerous, as it encourages the formation and discharge of static electricity. 'Air-conditioning' is a much misused word, since it means the control of humidity as well as DB temperature; and often extends also to efficient air filtration.

Humidity control is the result of a correct balance of air temperature and moisture content. This is provided automatically in ranges of close-control, packaged units of the same shapes and cooling capacities as those described previously, and which have so many applications that those of us who work on air-conditioning equipment are certain to install, maintain, and service them.

Room temperature

Room temperature is, as we have already seen, sensed and controlled by a room thermostat, which cycles a refrigeration system. Obviously, if we have several circuits within the system, we can control room temperature more closely than if there is only one. But in close-control applications, we are also very interested in *coil* temperature, because this determines how much moisture is removed from air passing through the evaporator. At a given air velocity and dew point the lower the coil temperature the more moisture is removed. Evaporators of close-control units are therefore designed and balanced with condensing units to operate at a predesigned suction temperature when used at design air-on-coil and air-onto-condenser conditions.

Obviously, actual internal and external air conditions vary from those used by equipment designers and this causes air to leave the evaporator at different DB and WB temperatures. If this results in the conditioned space becoming too dry, we must add moisture with a humidifier. If things go the other way, and the space relative humidity becomes unacceptably high, we must activate a *reheat coil* to reduce the RH. Both will be controlled by a humidistat.

Humidistats

Working on modern, close-control equipment will lead to your handling electronic, rather than electro-mechanical, thermostats and humidistats. Do not let this worry you – both produce the same end results and electronic controls require little maintenance, other than being kept scrupulously clean.

In an electro-mechanical control, the human hair or plastic humidity sensor changes in length if the RH of the air passing over it varies. This movement is transmitted through a lever mechanism to

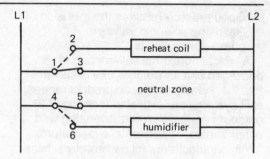

Fig. 7.3 Humidistat switching action

operate double-throw electric contacts. Figure 5.20 illustrated a typical switching action for a thermostat designed to select cooling or heating modes. Figure 7.3 shows an arrangement of contacts which would be operated to actuate one humidifier and one step of electric heating.

Referring to Fig. 7.3: if space humidity exceeds the set point, the contact between terminals 1 and 3 is broken, and that between 1 and 2 is made. This makes the control circuit of a reheat coil and switches it on. When space humidity falls to the set point, the circuit between 1 and 2 is broken, the reheat coil switched off, and the controls rest in their neutral zone. If humidity continues to fall, the circuit between terminals 4 and 5 is broken, and that between 4 and 6 is made. This switches on a humidifier, which operates until RH increases to the set point, contact 4–6 is broken, and neutral zone position 4–5 is again selected.

One stage of heating provides only coarse control and it is more usual to use, say, three steps of electric heating, with one heating element across each supply phase, together with a humidifier. In this case, an electronic sensor will probably be used. Details of the RH of air passing over a plastic sensing strip are transmitted to an electronic controller in which the signals are amplified and analysed and which governs the rotation of a reversing-type 24 V electric motor. This turns in one direction if RH rises, and the other if RH falls. The motor is coupled to a multi-step controller, the contacts of which are open when RH is at the set point. They close to switch on the humidifier if RH falls and close to switch on one or more stages of electric reheating – in proportion to the size of the load sensed – if RH rises. The contacts on this type of controller are usually adjustable to vary the number and band width of individual operating stages.

Factory settings are likely to be for a neutral zone width of 5 per cent RH, and operating differentials of 3 per cent. If de-energised, the motor automatically moves into the neutral position.

Reheat coils

Sheathed-type electric reheat coils are normally used in humidity control schemes. They are easy to control, and quick to reach operating temperatures – or to cool down when de-energised. Safety thermostats in each circuit safeguard against thermal overloads, and reduce fire hazards. Quality products incorporate indicating lamps, which show if and when each heating coil is energised. Most units can contain a number of coils, which are available in a broad range of capacities.

Humidifiers

Several different types of humidifier are available, all designed to admit steam or spray small water droplets into the supply airstream where the moisture will be vapourised and immediately increase RH. Again, a wide range of capacities is available, equipment being rated in terms of the moisture it will distribute, or the steam it will generate, in a given period of time – usually one hour. The types most often used are listed below.

Steam pan humidifiers
Water is gravity-fed into a closed container which contains one or more electric elements. If the control circuit is made, elements are energised and boil some of the water. The steam passes out of the boiling pan and is fed directly, or through a supply line connecting the pan and steam distributing nozzles, into the supply airstream which has left the evaporator. Water which has been boiled is replaced from a break tank fitted with a ball-type make-up valve. When space RH reaches the set point, the humidifier de-energises the heating elements.

This arrangement, illustrated in Fig. 7.5, is simple and economical. Drawbacks are that there is a time lag between the need for steam being sensed, and its being generated, and that in hard water areas heating elements will – like those in electric kettles – become coated with scale, which increases the time taken to produce steam. This can of course be removed using a standard descaling chemical, but if treatment is neglected the humidifier will

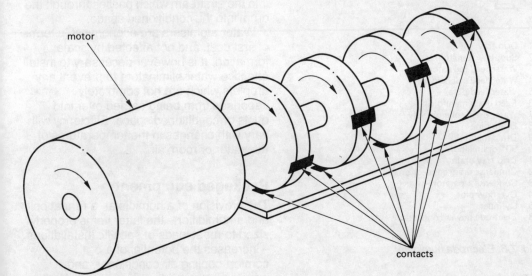

Fig. 7.4 Motorised step controller

TROPICAL REFRIGERATION AND AIR-CONDITIONING

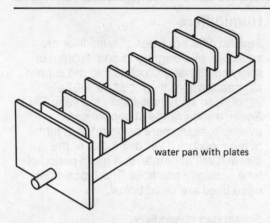

Fig. 7.5 Steam pan humidifier

cease to operate effectively. Water bleed-off or water drainage and refilling cycles initiated by time clocks are often provided, to reduce scaling.

Electrode humidifiers

The control sequences and water supply arrangements of this type are as described above, but steam is generated by the action of electrodes, so that there are no heating elements which can be scaled over by hard water. Scale does of course form as water boils, but accumulates in the container and takes much longer to affect the efficiency of the system. Sophisticated control systems can be used to adjust the current drawn according to the pH of the water and, where necessary, to boil or drain off and replace part or all of the water in the boiler until its chemical characteristics are suitable for the equipment.

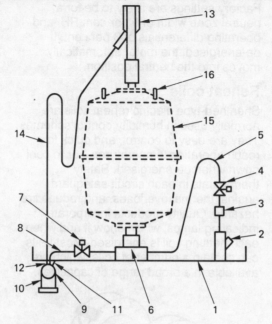

1 Drip tray
2 Supply water strainer
3 Water feed regulator
4 Water feed solenoid valve
5 Water feed pipe
6 Feed drain manifold
7 Water drain solenoid valve
8 Drain pipe
9 Drain manifold
10 90° drain elbow
11 Drip tray drain
12 Condense drain connection
13 Condense separator
14 Overflow pipe
15 Cylinder
16 Electrode power connections

Fig. 7.6 Electrode humidifier

Electrode humidifiers are more expensive than steam pan types, but offer advantages where water characteristics lead to heavy scale formation. The average time taken to produce steam is minimised by the absence of scale on heating elements.

Water atomising humidifiers

The spinning disc type shown in Fig. 7.7(a) uses a horizontal, quickly-rotating surface to atomise (make into tiny droplets) water sprayed onto it, or admitted in drops from an overhead supply grid. Rotary drum construction (see Fig. 7.7(b)) features a horizontal drum or brush construction, in which only its top portion rises above water level. As the drum rotates it sprays a shower of fine droplets into the airstream which passes through the drum into the conditioned space.

Water atomisers are economical in terms of first cost, and not affected by scale formation. It is however necessary to install effective water eliminators to prevent any droplets which are not completely vaporised from being carried over into ducts or conditioned space. Efficiency will vary with changes in the temperatures of the water or room air.

Packaged equipment

The provision of a humidistat, a reheat coil, and a humidifier – the latter items properly sized to meet loads of specific installations – increases the potential of a comfort-cooling air-conditioner, and enables it to provide close control. Air filtration efficiency must be above normal

COMMERCIAL AND INDUSTRIAL AIR-CONDITIONING EQUIPMENT

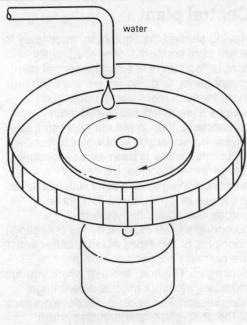

a) Spinning disc

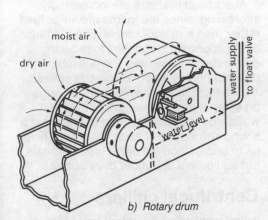

b) Rotary drum

Fig. 7.7 Water atomising humidifiers.

standards in the majority of installations requiring close humidity control.

Packaged units incorporating these additional items of equipment, factory-wired and tested, are available from specialist manufacturers. In addition to normal vertical arrangement of the airhandling components – the fan section above cooling coils and air filters – a layout which allows downward discharge is often required in computer suites or other applications where cooled and humidity-controlled air must be drawn into the base of electronic equipment from underfloor supply ducts or plenum chambers. Airhandling units meeting both requirements are available as standard. Cooling coil options include direct-expansion or chilled-water types. Air-cooled and water-cooled condensers, remote condensing units or glycol-cooled condensers are also readily available.

There need be little external difference between comfort- and close-control equipment, as illustrated in Fig. 7.8, other than the presence of indicating lamps on the more complex control panels of the close-control models.

Equipment wiring diagrams provide a clearer indication of the differences; and Fig. 7.9 provides comparisons based on single-circuit, direct-expansion airhandlers. Closer scrutiny of Fig. 7.9(b) shows that the humidity control equipment can operate independently of the refrigeration system, although (as stated in the introduction to humidity control) evaporator temperature is of vital importance in maintaining correct humidities in an economical manner.

a) Comfort-cooling model

b) Close-control model

Fig. 7.8 Packaged air-conditioners

115

TROPICAL REFRIGERATION AND AIR-CONDITIONING

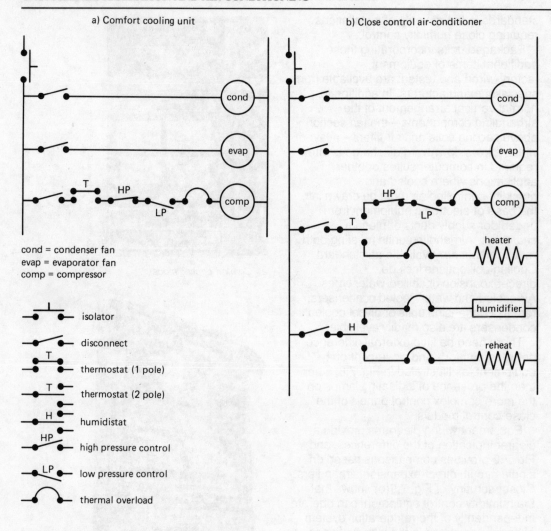

Fig. 7.9 Wiring diagrams

Central plant

Having studied the equipment necessary to adapt from comfort cooling to humidity control, we can now examine central plant components. In doing so, we shall find that there is much more use of packaged, factory-assembled and tested main components than in the refrigeration trade. This is particularly true of water chillers – and chilled water is used as a secondary refrigerant in the great majority of air-conditioning installations with capacities in excess of 210 kJ/s (60 TR). Indeed, before looking at other system components, we must study the operating principles of two types of large chiller which are normally supplied completely packaged. The first, and that which you are most likely to work on, has a centrifugal compressor. The second is a development of the absorption refrigeration system examined in Chapter 5.

Absorption systems are increasingly interesting, since the 'prime mover' is heat, rather than a motor or an engine. Because of the importance attached to reducing demands for fossil fuels there is much development work on absorption chillers, and air-conditioners which can run largely on solar energy. These are not yet in worldwide use, but illustrate the fact that a sound knowledge of basic theory will help us to understand not only current models, but equipment still under development.

Centrifugal chillers

When discussing the rotary compressors used in domestic appliances we said that

COMMERCIAL AND INDUSTRIAL AIR-CONDITIONING EQUIPMENT

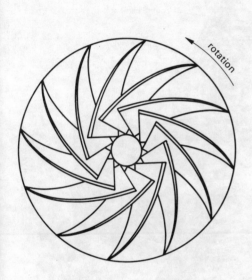

Fig. 7.10 *Centrifugal compressor impeller*

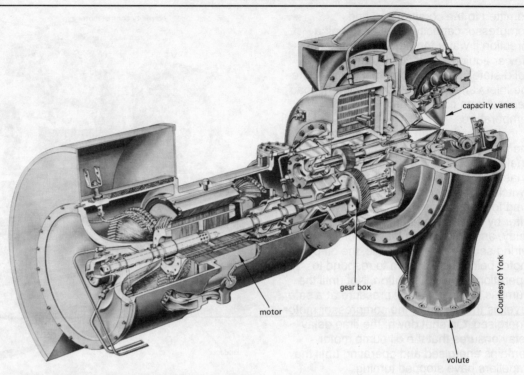

Fig. 7.11 *Cross-section of centrifugal compressor*

they are not well suited to high pressure work, and often use a specially developed low pressure refrigerant. The large centrifugals in current use do not rely upon rolling pistons or rotating vanes. Their impellers have been developed from turbines, and use fixed blades to help compress refrigerant vapour thrown outwards by high centrifugal forces. A typical impeller is shown in Fig. 7.10. High speeds are common because centrifugal compressors do not have as high a compression ratio as reciprocating or screw designs. Two-pole motor speed, 3 600 r.p.m. on 60 Hz, is described as 'slow' and gear boxes are used to increase r.p.m. to as much as ten times that figure. Alternatively we can use more than one stage of compression to enable them to operate efficiently over a realistic range of (suction to condensing) pressures when using R11 or R12. Two-stage compressors are frequently used, the discharge port of the first stage delivering refrigerant to the inlet port of the second.

Figure 7.11 illustrates the construction of a single-stage compressor, and has been chosen because the section shows the capacity control system most used in this type – inlet guide vanes. The positions of vanes can be varied from fully open to fully closed, by a pneumatic or electric motor actuated by a mechanism responding to a chilled-water temperature sensor. Older designs probably have the sensor at the chilled-water outlet, working through a pneumatic motor to move the vanes in a closing direction if water temperatures fall (thus reducing the amount of refrigerant admitted to the compressor, and compressor capacity) or in an opening direction if water temperatures increase. Newer equipment is likely to rely on solid-state, electronic controls to modulate the inlet vanes in accordance with the load sensed from chilled-water inlet temperatures.

In view of the high running speeds, and the fact that this type of compressor is very finely balanced and takes a long time to coast to a standstill when its motor is switched off, particular attention must be paid to lubrication. This is normally looked after by a motorised pump, a time delay relay, and an oil pressure control. The wiring sequence is automatic, the oil-pump motor being the first item to respond to operation of the start button. Not until the pump is running, and oil pressure at a safe level, is the starter of the compressor motor energised. On shut down, the time delay relay ensures that the oil pump motor remains energised and operating until the impellers have stopped turning.

Figure 7.12 is a cutaway illustration of a complete, packaged, centrifugal chiller. The machine is a hermetic, single-stage design, with the electric motor and gearbox at the top left. The suction line is at the top right, with compressor and discharge *volute* between it and the motor. The mechanism above the control panel is the capacity control motor, linked to inlet vanes as already described. The shell-and-tube condenser can be seen beneath the compressor volute connection to the discharge line, with a shell-and-tube evaporator located below and behind the condenser, and both are connected to

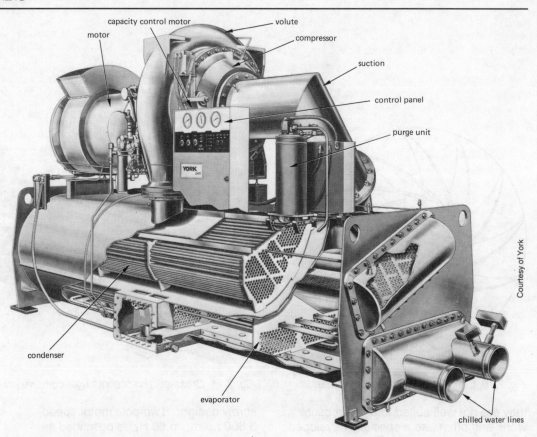

Fig. 7.12 Cutaway of centrifugal chiller

water *headers*. If in doubt as to which heat exchanger of a water-cooled chiller is which, base your checks on the assumption that the evaporator will be the larger, and lower, of the two shells or header boxes.

Comments on the application of air-cooled condensers to larger equipment, made in Chapter 6, are applicable to centrifugal chillers. Factory-assembled air-cooled units are available in sizes up to 1 230 kJ/s (350 TR) and larger sizes are assembled on site.

One final point of detail on centrifugals applies also to absorption-type chillers – the need for purge units. When low

COMMERCIAL AND INDUSTRIAL AIR-CONDITIONING EQUIPMENT

pressure refrigerants (e.g. R11) are used in centrifugals, suction pressures are usually lower than that of the atmosphere. Similarly, in absorption circuits, part of the system operates at very low pressure. The equipment sizes are so large that it is extremely difficult to prevent air entering low sides through non-welded joints; and as always, air contains moisture. The mixture of (non-condensable) gas and water vapour accumulates in the condenser, from which it is automatically vented by the purge unit.

Principles of operation can be explained with the help of Fig. 7.13. A mixture of refrigerant and contaminants is drawn from the top of the condenser, and compressed by the purge unit compressor. This discharges into an oil separator, from which oil is returned to the compressor. A heater is used to help separate it from the vapour. The warm vapour passes into the purge drum, which contains a finned, chilled-water cooling coil. This causes the refrigerant vapour to condense, and collect at the bottom of the drum. A float valve enables the refrigerant to return to the evaporator.

The non-condensable gases remain at the top of the drum, and as their pressure increases a pressure relief valve (see Fig. 6.18) opens and vents the air to atmosphere. Any accumulated water will float on the surface of the refrigerant in the drum, and can be drained off through a manually-operated valve.

Purge units operate automatically (operating and safety controls of centrifugal and absorption chillers are invariably factory-wired and operate in the correct

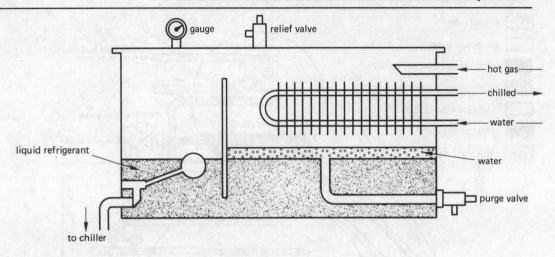

Fig. 7.13 Purge unit

sequence following the use of either the start or the stop button) but need routine checks and maintenance. They are not normally found in centrifugal systems using R12, which do not operate in a vacuum.

Absorption water chillers

The absorption system used in domestic refrigerators, and illustrated in Fig. 5.16, uses ammonia as the refrigerant, and water as the absorber in which refrigerant is circulated before being vapourised by heat from the boiler. In large chillers, it is necessary to use different fluids in a roughly similar manner. The refrigerant is distilled water. (The operating pressure of the system is one in which water boils at 4·4 °C (40 °F). Water, as we noted when studying basic theory, has a high latent heat of vaporisation.) The absorbent is generally a solution of lithium bromide in water. It is extremely 'thirsty' for more water and a little absorbent can contain a lot of refrigerant.

Figure 7.14 illustrates, schematically, a typical lithium bromide/water chilling circuit. In the water-cooled condenser, refrigerant (water) vapour at low pressure is cooled, and condenses. The liquid refrigerant then flows into the evaporator, where it is pumped over heat exchange surfaces and vapourises, removing heat from chilled water flowing through the heat exchanger tubes. The refrigerant vapour is drawn into the absorber – the part of the system at the lowest pressure. Here it is cooled by a coil containing condenser water, and reverts to its liquid form. That liquid – water, remember – is greedily absorbed by the lithium bromide. The resultant solution is pumped via a heat exchanger to the high

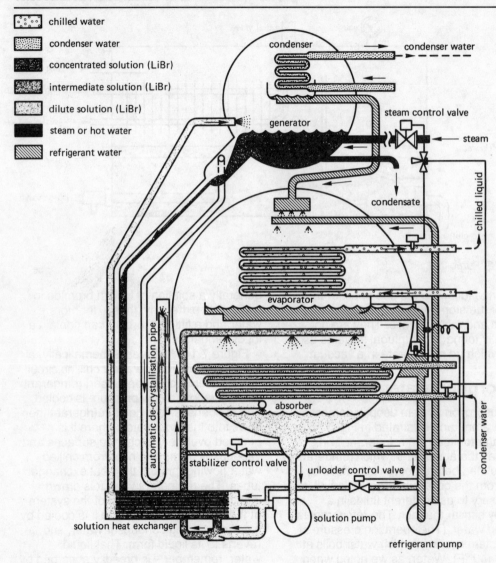

Fig. 7.14 Absorption chiller circuit

pressure side of the system – the generator (which we previously called the boiler). At this point the mixture is heated by steam, or high-pressure hot water. The refrigerant (water) vapourises, and enters the condenser to start another cycle. The lithium bromide solution returns to the absorber, and waits to collect more liquid refrigerant.

Chillers of this type are available in sizes up to 5 275 kJ/s (1 500 TR) and offer several advantages. The only moving parts are those in pumps and purge unit and the only electricity needed is that required by pumps and controls. Operation is silent. On the other hand, the equipment is very large and heavy and operating efficiency falls quite sharply if condenser water is not available at relatively low temperatures. This is the reason why absorption chillers are not so widely used as centrifugals in humid tropical areas.

As we have already seen, purge systems are needed to remove non-condensable gases. In this case, they enter a pick-up tube in the absorber and pass into a purge chamber, which is at low pressure. Water vapour and non-condensable gases are here released into a spray of absorbent which removes the water and flows from the base of the chamber, back to the absorber. The remaining non-condensable gases are removed by a vacuum pump, which discharges them to atmosphere. A schematic sketch of this part of the system appears as Fig. 7.15.

Absorption chillers are made to several different designs which affect operating details. The basic principles quoted above can be taken as representative, but

COMMERCIAL AND INDUSTRIAL AIR-CONDITIONING EQUIPMENT

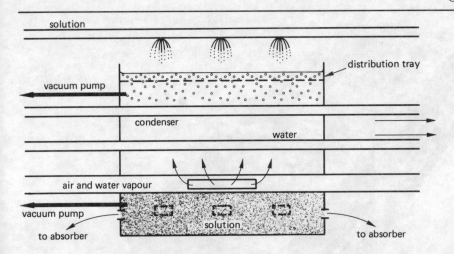

Fig. 7.15 Absorption chiller purge chamber

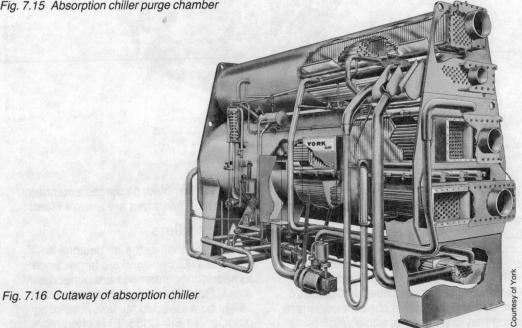

Fig. 7.16 Cutaway of absorption chiller

manufacturers' literature must be used to ascertain detailed operating and maintenance needs. A cutaway view of a typical packaged chiller is included (Fig. 7.16) to enable students to visualise the complex nature of the various circuits and heat exchangers which it contains.

Packaged water chillers

Centrifugal compressors and absorption systems are only two of a range of packaged, water-chiller options, which are extensively used in air-conditioning applications. For all practical purposes, direct expansion installations cease to be a practical proposition in jobs much over 700 kJ/s (200 TR) and even here the buildings concerned need to be compact (cinemas, for example) to avoid excessive pipe lengths, and pressure drops. The best alternative, used in the majority of major installations, is chilled water.

Compressors

The types available have all been reviewed earlier – welded hermetics up to 5·6 kW ($7\frac{1}{2}$ hp) are often used in multiples with independent refrigeration circuits, on jobs up to eight times that size. From 7·5 kW (10 hp) to 150 kw (200 hp) accessible hermetic, reciprocating units are currently most popular. Their compressors all incorporate capacity control, and part wind or equivalent reduced voltage starting devices. The screw compressor is not yet as popular – it is used more for refrigeration than air-conditioning; and from 75 to 750 kW (100 to 1 000 hp) centrifugal compressors are most popular. As already

121

noted, absorption systems are less common in the tropics, as condenser water is rarely available at entering temperatures below 32 °C (90 °F). Accessible hermetic construction is favoured for centrifugal and screw designs, which must include reduced voltage starting, and capacity control. In a few rare cases, 11 000 volt motors are used in very large centrifugals, to avoid the need for transformers; but these should only be handled by specially trained personnel.

Condensers

Air-cooled

Structural details are as described in Chapter 6, and the use of motor speed controllers or fan outlet dampers is more common in tropical installations than would at first sight seem likely. It must be remembered that condensers operating in extremely hot climates are 'oversized' to secure acceptable condensing pressures by day in extreme heat. When temperatures fall overnight, or in winter, condensers may become effectively oversized and be unable to maintain liquid refrigerant at pressures high enough to ensure satisfactory system operation. Whilst it is not necessary to build solar screens for properly designed condensers, it is desirable to take advantage of any shade provided by buildings when outdoor temperatures are at their peak. Vertical air-discharge, or the use of centrifugal fans to handle the approximately 110 ltr/s of air per kJ/s (800 cfm per 1 TR) without undue noise, is common.

Fig. 7.17 Air-cooled packaged chiller

Water-cooled

Apart from systems with welded hermetic compressors (which generally use tube-in-tube, or tube-in-shell condensers because they are cheapest) the use of shell-and-tube models is standard in units with 7·5 kW (10 hp) or larger motors. Construction is as described in Chapter 6. The use of evaporative condensers is rare, since they are not compatible with factory-assembled, packaged equipment. Pressure relief valves are always fitted.

Water chillers

The use of shell-and-tube designs is invariable. Dimensions are large, since temperature differences are of necessity small. Refrigerant temperatures cannot safely be allowed to fall below freezing point, and 1·66 °C (35 °F) is often regarded

COMMERCIAL AND INDUSTRIAL AIR-CONDITIONING EQUIPMENT

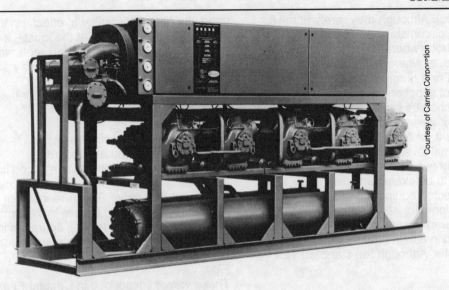

Fig. 7.18 Water-cooled packaged chiller

as the limit; whilst air-off-coil conditions at the secondary evaporators, are generally of the order of 15·5 °C (60 °F) to secure adequate cooling and dehumidification. The usual temperature limits for chilled water are: entry at 14·4 °C (58 °F) and discharge at 5·5 °C (42 °F). Lower temperature ranges require the use of glycol solutions. Good thermal insulation is essential if condensation is to be avoided. It normally takes the form of 50 mm (2 ins) of foamed insulation material protected by vapour-tight sheet-metal covers. In both chillers and condensers, water pressure drops must be kept to a low level, to avoid wasting money and energy on needlessly large pumps and piping circuits. Clear details of pressure drops should be quoted in manufacturers' literature, but as a rule of thumb, 60 kPa (20 ft of water) is nearing the maximum acceptable.

A factor which clearly cannot be controlled by equipment manufacturers, but has a major effect on the performance of water chillers or water-cooled condensers, is scaling. If scale is allowed to form, or other impurities to build up on heat exchanger surfaces, performance must inevitably suffer. Nearly all equipment ratings are based on a fouling factor of 0·1 $m^2.°C/kW$ (0·0005 $ft^2.°F.hr/Btu$) in water-cooled condensers; but under realistic site conditions, recirculating normal water, it is difficult to prevent the fouling factor actually secured from rising to double the figure just quoted. This has, in practice, the effect of increasing chilled water temperatures by about 1 °C (2 °F); and it is essential that condensers are properly sized, and water continuously treated, to enable design temperatures to be secured. The problem is much less acute in the case of water chillers, since chilled water should be both chemically treated and contained in an airtight system, so that there is comparatively little chance of scale formation or other insulation of the heat exchanger surfaces.

Instrumentation

It is good practice to supply control panels including dial type instruments showing water inlet and outlet temperatures at each heat exchanger and suction and discharge refrigerant pressures with all chillers with capacities above 176 kJ/s (50 TR). An indication of the current drawn by each large motor is desirable, ammeters being fitted at motor control centres, or on starter cabinets.

Operating and safety controls

The basic essentials are: high and low pressure refrigerant controls and oil-pressure safety controls to suit the type of compressor. A water-temperature safety control is invariably fitted in addition to the operating thermostat, and it is desirable that large chillers are also protected against freeze-ups by water flow switches. These have paddle-type sensors to detect water movement, which make the circuit – and allow current to flow through the system controls – only if the pressure of water circulating through the system is at a safe level. If the volume of water circulated falls sharply – due perhaps to the failure of a water pump or the accumulation of scale

TROPICAL REFRIGERATION AND AIR-CONDITIONING

in a strainer – the flow switch breaks the circuit, and closes down the entire system.

Electrical safety controls provided by the factory do not normally include more than high quality temperature sensors buried in motor windings and the interlocking of sequence switches to prevent compressors, for example, from starting until oil pressure is up to a safe level, condenser fans or pumps are running and chilled water pumps are running and circulating enough water to satisfy the flow switch. Electricity suppliers are normally legally bound to supply power within tolerances of plus or minus 6 per cent of nominal voltages, and 1 per cent of frequency (for example, 49 to 51 cycles/sec). Although they rarely admit it, there are circumstances – often beyond their control – where these tolerances are exceeded, or where a phase is 'lost' entirely. For a number of areas, therefore, equipment suppliers may include in their control schemes sensors which close down the plant if voltages deviate by more than 10 per cent (or any other desired value) from normal levels, or if a phase is lost, or even reversed. The wiring diagram is the first item to be examined when large, well-protected equipment is being commissioned or serviced!

Refrigerant controls can differ from those described in earlier chapters and paragraphs. Although the standard TEV is used whenever possible – in effect, up to about 352 kJ/s (100 TR) at the most – it is not physically able to accurately meter the volumes of refrigerant required by say a 880 or 2 640 kJ/s (250 or 750 TR) evaporator. The equipment designer then has the following two options.

Pilot-operated expansion valves
A solenoid valve and a standard thermostatic expansion valve in a loop line are used to control a piston-operated, main refrigerant flow valve. This arrangement is particularly well suited for use in systems with capacity control devices operating in marked steps (such as cylinder unloaders).

Float valves
These valves (described in Chapter 6) are used in chillers handling very large volumes of liquid refrigerant – such as a 1 055–1 410 kJ/s (300–400 TR) centrifugal charged with R11. In this type and size of plant, the use of flooded evaporators is common, and in many instances float chambers are built into the evaporator shells to secure maximum efficiency.

Other chillers

The use of submerged evaporators in insulated tanks of chilled water, available to be pumped around airhandlers as required, which can even out load fluctuations, is uncommon in air-conditioning applications. Similarly, the use of 'ice banks' in which water surrounding evaporator pipes is deliberately frozen, to store up the latent heat of accumulated ice and so enable

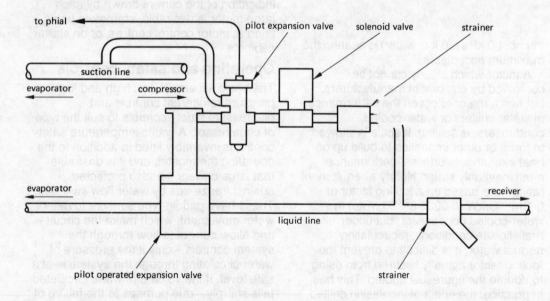

Fig. 7.19 Pilot-operated expansion valve layout

COMMERCIAL AND INDUSTRIAL AIR-CONDITIONING EQUIPMENT

compressor capacities to be kept to a minimum, is normally limited to refrigeration applications such as milk cooling over Baudelot coolers.

All chiller types described in this chapter are widely used in industrial processes, chilled water often being used to remove heat from moulds or dies used in production machinery. Some industrial requirements necessitate water being available at steady temperatures, which may require heating as well as cooling, and it is in this type of application that storage tanks are most often used. Thermal control can be effected by jacketed electric heating coils and bare or finned tube chilled-water coils inserted in an insulated tank. Heating or cooling functions are controlled by thermostats with a neutral zone between the switching contacts for the two functions. Alternatively, a tank of chilled water and a tank of warmed water can be individually supplied and controlled, but have their contents mixed by a modulating-type three-way valve to achieve the set point temperature.

Drinking water coolers

These are usually regarded as domestic appliances, and are described in chapters devoted only to domestic products. We have in this instance included them in the overall family group of water chillers, since they are comparatively rarely used as items of residential equipment in the Tropics. This stems not only from cost considerations, but from the purity and pressure of water available direct from distribution mains.

These considerations apply also in some areas outside the tropics, and result in the production of bottle-type coolers, as well as 'pressure' types for connection to the mains. Capacities range from a nominal 0·003 to 0·03 ltr/s (3 to 30 gallons per hour), with bottle types normally limited to a maximum of 0·008 ltr/s (8 gph) by the weight of water and resulting handling difficulties. Equipment ratings should be based on an ambient temperature of 32·2 °C (90 °F) and a water cooling range of 26·7 °C to 10 °C (80 °F to 50 °F) for pressure types, or 32·2 °C to 10 °C (90 °F to 50 °F) for bottle models.

For obvious reasons, drinking water is normally contained in stainless steel tanks and distribution pipes, with cooling coils either wound around the exteriors of tanks, or embossed into the outer surfaces in similar fashion to the cooling coils in domestic refrigerators (see Fig. 7.20).

Refrigeration systems invariably incorporate welded hermetic compressors with single-phase electric motors; and forced-draught, cross-finned condensers. Capillary tubes are used almost exclusively; R12 is the most common refrigerant and the diverse operating and safety controls used on commercial and industrial-size chillers are omitted because of their cost. Only a water thermostat is normally retained. All pressure-type coolers must contain, or be plumbed in to include, properly trapped drains.

When pressure-type coolers are of the fountain type, with chilled water being supplied in a low jet to which the lips are lowered (as opposed to being fitted with valves from which water flows into a glass

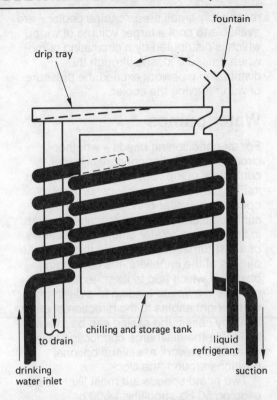

Fig. 7.20 Cross-section of pressure-type water cooler

or cup) an appreciable percentage of all water chilled is spilled, and runs to drain. The generally accepted water flow rate is 0·03 ltr/s (0·5 usgpm), a good half of which is normally lost. Better product designs therefore require this wasted water to drain through a coil which precools incoming water on its way to the cooling/storage tank. Where numerous supply points are required within a

TROPICAL REFRIGERATION AND AIR-CONDITIONING

reasonably small area, 'central coolers' are available to cool a larger volume of water, which is distributed by a circulating pump when pressure losses through the distribution pipework exceed the pressure of water leaving the cooler.

Water pumps

For air-conditioning needs – whether circulating chilled or condenser water – centrifugal pumps such as the one shown in Fig. 7.21 are invariably used. These operate on similar principles to the centrifugal compressors described earlier in this chapter. Water enters at the centre of a fast-turning impeller; it is thrown to the outside of the impellers, the converging blades of which add to its speed; and it collects in a volute leading to a discharge port at right angles to the direction of entry. In many cases, the casing can be rotated to enable the discharge connection to be made to pipework at several optional positions around 'the clock'.

Two pump speeds are most likely to be used on 50 Hz supplies: 1 500 or 3 000 r.p.m. synchronous. The faster the speed, the greater the head against which the pump can deliver water, but the more noise it generates. There is also a risk that pumps running at the same speed as centrifugal compressors may set up sympathetic vibrations which affect both, so the lower running speed is usually favoured.

The bodies of pumps designed to handle water are likely to be of cast-iron construction, trimmed in bronze with a bronze or stainless-steel shaft, and bronze

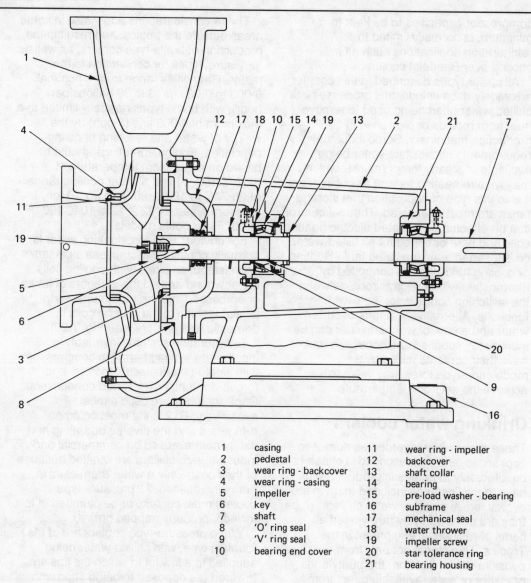

1	casing	11	wear ring - impeller
2	pedestal	12	backcover
3	wear ring - backcover	13	shaft collar
4	wear ring - casing	14	bearing
5	impeller	15	pre-load washer - bearing
6	key	16	subframe
7	shaft	17	mechanical seal
8	'O' ring seal	18	water thrower
9	'V' ring seal	19	impeller screw
10	bearing end cover	20	star tolerance ring
		21	bearing housing

Fig. 7.21 Centrifugal water pump

impeller(s) – like compressors, pumps can have several stages to enable them to operate against higher pressures – and bearings. Water leaks are normally avoided by the use of rotary shaft seals. The old fashioned stuffing box is rarely seen on new equipment.

There are several means of connecting the motor and the pump. If the pump is mounted on a baseplate a direct-drive coupling is usually employed, but some prefer to use pulleys and vee belts. The most foolproof arrangement is that used in accessible hermetic compressors with both pump and motor on a common shaft and secured within a single-piece or close-coupled body assembly.

Whilst installation and instrumentation needs are covered in Chapter 8, it should be noted that pumps should always have a valve, an anti-vibration coupling and provision for the use of a pressure gauge on both suction and discharge sides. The use of a water strainer immediately before the pump is essential. In many cases, a second pump is piped into the circuit ready for use if the first fails. This entails some complicated pipework, and, in order to avoid the need for this, an increasing number of manufacturers now offer factory-assembled dual-pump assemblies, with both the running and standby pumps completely plumbed in at the factory. One such assembly is shown in Fig. 7.22.

There is little basic difference between pumps designed to handle water, brine or refrigerants. All components in contact with the liquid being handled must be chemically compatible with it at operating temperatures, but operating principles are little changed.

COMMERCIAL AND INDUSTRIAL AIR-CONDITIONING EQUIPMENT

Airhandling units

The main duties of any airhandling unit, whether a factory package or field-assembled, must be capable of some or all of the following.

a) Mixing return and ventilation air supplies.
b) Filtering all air to required standards of efficiency.
c) Cooling, or cooling and dehumidification, with or without reheating the total airstream **or** heating, or heating and humidification.
d) Avoiding carrying over into ductwork or conditioned space, any moisture deposited on cooling coils during the dehumidification processes or by water

Fig. 7.22 Dual pump assembly

Fig. 7.23 Central plant airhandling unit

127

sprays, or fed into the airstream by humidifiers.

e) Discharging all air at a pressure equalling the sum of the resistances to airflow offered by ductwork, supply grilles and diffusers, and all airhandling unit components.

Each of these requirements can be satisfied in one of several ways. We shall note most in passing, but study in detail only those in widespread use in our specific (geographical) areas of interest.

Air mixing

Fresh, ventilation air is normally drawn into plant-rooms through weatherproof louvred intakes fitted with volume dampers and mesh screens, which prevent the entry of leaves, birds and insects. It is filtered before being mixed with return air in either a plenum, or a mixing box fitted with opposed-blade dampers to regulate the volume of each type. In regions with wide seasonal variations in temperature, air intake volumes are often varied by thermostatic controls reacting to changes in external temperatures, to reduce demands for refrigeration or heating at times when the ambient air temperatures approach design air-off-coil levels.

Filtration

There are three types of filter for normal use:

a) Throwaway types, of foamed plastic or glass fibre construction.
b) Cleanable types, normally using metal 'wool' or fibre as the filtering medium. This can be washed clean, and may be coated with oil or other viscous material to increase its efficiency.
c) High efficiency filters, used dry and made from pleated mats of wool felt, cellulose fibre, or synthetic materials. The choice of material in this and other types of filter depends upon the efficiency required, and the velocity of the air to be filtered.

Efficiency ratings are usually expressed in terms of the percentage of air contaminants of a specified size, removed when the air passes through the filter at a known velocity. For example, 95 per cent at 5 micron is a good standard filter performance; with a normal high efficiency filter rating being 80 per cent at 0·5 micron; both at a velocity of 2·54 m/s (500 ft/min). Where high standards of air cleanliness must be maintained, differential pressure gauges can be used to sense increased resistance to airflow resulting from the accumulation of dirt on filters. This can be read directly, or the pressure change used to operate a signal to change the filter when resistance reaches a preset level. Automatic roll filters which react to the same information by automatically winding on a new section of filter material, are very effective, but their use is limited by first costs. Electrostatic or electronic air filters are discussed later.

Cooling coils

All air-cooling or heating coils are of copper-tube, aluminium-fin construction, with spacing options varying from 6 to 14 fins per 25 mm. Fins are often corrugated to increase the turbulence of air passing through the coil, and thus increase its heat transfer capacity. Moisture carry-over is prevented by limiting air velocity to approximately 3·0 m/s (600 ft/min).

Direct-expansion coils

These must be in refrigeration-quality copper tubing with formed return bends (preferably silver-soldered) and guaranteed for pressures of 2 410 to 2 760 kPa (350–400 psi). Evaporators of this type (and also air-cooled condensers) are normally tested for leaks when first manufactured. This is done by charging them with dry nitrogen at the required pressure, completely immersing them in back-lit, coloured water and detergent tanks and looking for escaping gas. This does not prevent the possibility of leaks developing later, following mechanical or other damage. Measurements for copper tubing are at present given in British units and $\frac{3}{8}$ in to $\frac{5}{8}$ in outside diameter (OD) are the most used tube sizes, depending upon the capacity of the coil. Refrigerant flow is evenly balanced over the tubes by a distributor located immediately after the TEV.

Coils are often divided into two or more separate circuits, each with its own TEV and distributor. Coil divisions can be made in the vertical plane (face control) or in depth (depth control). Depth control might, for example, use two rows of coil for each of two circuits. This approach requires fine balancing, as the first two or three rows of a coil will always remove more heat than the next two or three, if both sections are at identical refrigerant temperatures. TEV's must therefore be set to match the design cooling capacity of each section. This layout has the advantage that the entire

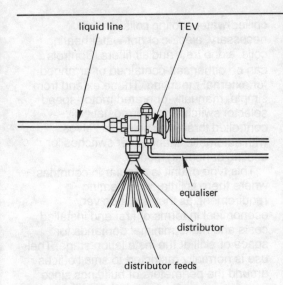

Fig. 7.24 Refrigerant distributor layout

face of the coil is always at a suitable refrigerant temperature. It provides better humidity control than a face-controlled coil, when capacity controls interrupt the supply of refrigerant to one complete section.

Chilled-water coils

The main difference between direct-expansion and chilled-water coils is that water designs use only one circuit. Capacity control is provided by modulating the volume of chilled water passing through the coil, or by using face and bypass dampers to vary the quantity of air passing over it.

The use of one continuous tube would result in excessive water pressure drops and, to avoid this, headered construction is used to supply a number of circuits in parallel. Air vent plugs are usually provided at the top of coils, and drainage plugs at low level. Water inlets are normally at the top of coils.

Condensate disposal

Cooling coils – and control valves – must be provided with full-size condensate trays, to collect moisture removed from conditioned air. These trays are normally of pressed-steel construction, insulated with sprayed-on polystyrene or polyurethane to prevent condensation forming on external surfaces. They also include provision for the connection of drainage lines.

Heating coils

Electric heating coil construction, and safety provisions, have already been described. We must also be familiar with hot-water and steam heating coils; which are sometimes needed in even the hottest of climates, to treat air used in industrial processes.

General construction methods are similar to those used for chilled water, but coil depth is only one or two rows and headered construction not therefore necessary. These coils are designated for use with a specific heat medium.

Low pressure hot water (LPHW)

The maximum temperature of water leaving a low pressure boiler is 121 °C (250 °F). Typical water conditions through coils are: entry at 82 °C (180 °F) and exit at 71 °C (160 °F). Air quantities over LPHW coils normally produce velocities between 2·0 and 4·0 m/s (400 and 800 ft/min).

Medium pressure hot water (MPHW)

This type, which is seldom used, has a maximum temperature of water leaving the boiler only slightly higher than that of LPHW types. In practice, water temperatures at coils are generally of the order of 127 °C (260 °F) at entry and 99 °C (210 °F) at discharge.

Steam

There is a wide variety of operating ranges of steam boilers and gauge pressures at coil entries can be between 34 kPa (5 psig) and 1 380 kPa (200 psig). These pressures are equivalent to temperatures of 108 °C (227 °F) and 198 °C (388 °F). Face velocities of air usually fall in the 2·54 to 6·60 m/s (500 to 1 300 ft/min) range. Coil tubes are normally larger (25 mm (1") OD) than those used for water, and installations must include steam traps and condensate return lines. Capacity control is by hot-water or steam flow-control valves. Water valves are likely to be of three-way, bypass design, and steam valves two-connection throttling types in supply lines.

Fans

It is unusual to find airhandling units of factory-assembled, or 'modular' types, which do not use centrifugal fans. These are normally double-width, double-inlet types with forward-inclined blades; and consist of scroll(s), impeller(s) and cone(s), and bearing support pedestals.
Sleeve-type bearings are normally used when motor sizes do not exceed 2·25kW (3 hp), plummer blocks being fitted for larger capacities. Fan speeds can normally be varied by altering the pitch of a motor

pulley designed for use with either one or two vee belts (although the use of direct-drive, variable-speed electric motors is likely to increase as electronic controllers become more competitive). The main components of a centrifugal fan, and the varying names which they may be called, are illustrated in Fig. 7.25.

Other types of centrifugal fans available have backward-inclined, radial or airfoil blades. The main advantages of centrifugal fans as a class are compact size, reasonable efficiency and low noise levels in applications against head pressures up to approximately 250 mm (10 in) *water gauge* (wg).

Cabinet types

Packaged airhandlers are available in vertical and horizontal designs, with vertical or horizontal air discharge. Fan sections should include, or have facilities for the easy connection of, flexible anti-vibration connections between fan discharges and duct connections. Cabinet construction methods normally require differences between those used for low and medium pressure applications 6 to 120 mm wg ($\frac{1}{4}$ to $4\frac{3}{4}$ in wg) and high pressure applications 125 to 250 mm wg (5 to 10 in wg). All cabinets, whether floor-standing or suspended from ceilings, should be isolated from building structures by spring or bonded-rubber type anti-vibration mountings. Cooling and/or heating coil connections should be made with flexible connections to prevent the transmission of noise or vibration through external pipework.

Other air terminal units

Smaller air terminal units are available in several main types.

Fan/coil units

These are available in vertical or horizontal designs, either in decorative cabinets or as basic units, to be 'furred in' behind structural or decorative features. They contain centrifugal fans with small direct-drive, single-phase electric motors, chilled-water cooling coils, and, if necessary, electric or hot-water heating coils, a drip tray, and air filters. Controls can be either self-contained or arranged for external mounting. These extend from simple, manually-operated motor-speed selector switches to thermostatically-controlled three-way solenoid valves with automatic changeover switches for cooling or heating modes.

This type of unit is popular in countries where there is little or no heating requirement, as they are then very economical in terms of first and installed costs and make minimal demands for space or skill at the installation stage. Their use is normally restricted to small offices around the perimeters of buildings since the maximum effective air-throw distance is approximately 5·5 m (18 ft).

High velocity induction units

These units dispense with fans, being supplied with high pressure, high velocity air from a central airhandling unit. This primary air is distributed from a central plantroom through high pressure ductwork (normally of circular, spiral wound construction) and discharged into individual units at velocities of 10·6–22·87 m/s (2 000–4 500 ft/min). On leaving the nozzles of the terminal units, fast-moving airstreams induce into their flow four or five times their own volumes of room air which passes through a secondary coil before being discharged into the conditioned space. The dew point of the primary air supply is designed to satisfy a little sensible and all latent heat loads. The fact

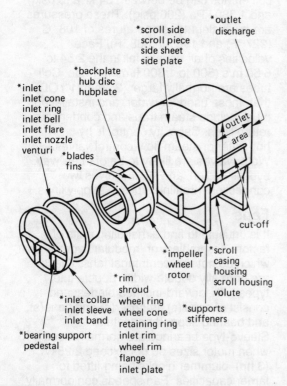

Fig. 7.25 Centrifugal fan components

COMMERCIAL AND INDUSTRIAL AIR-CONDITIONING EQUIPMENT

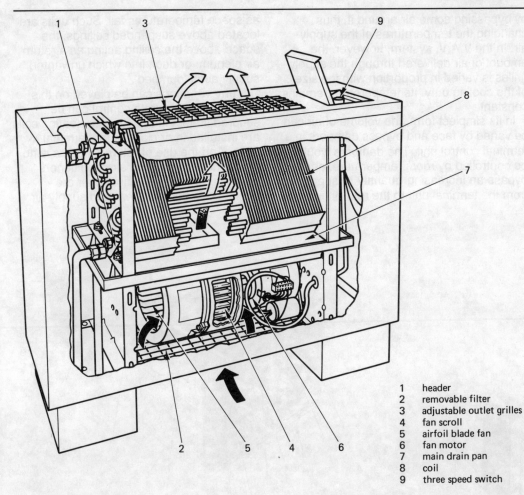

1 header
2 removable filter
3 adjustable outlet grilles
4 fan scroll
5 airfoil blade fan
6 fan motor
7 main drain pan
8 coil
9 three speed switch

Fig. 7.26 Fan/coil unit

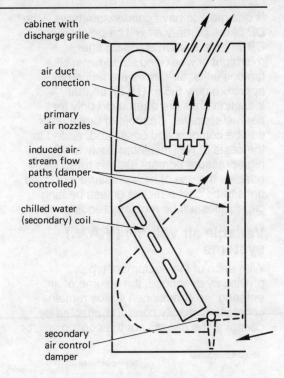

Fig. 7.27 Induction unit section

that there is no use of return air at the main plant makes it necessary that primary air volumes meet all ventilation requirements. The secondary coil is designed to take care of variations in sensible heat which result from changes in ambient temperatures. In tropical areas, there is invariably only a cooling requirement, and secondary coils can be supplied with chilled water at all times. In regions where heating may be required, a hot water supply must be available. There is a change-over condition at which the correct water supply must be used at the secondary coil; and further capacity control can be provided by a modulating air damper which regulates the amount of induced room air which can pass over the secondary coil.

All units need air filters to remove dust or dirt from induced air; and in tropical areas it

131

TROPICAL REFRIGERATION AND AIR-CONDITIONING

is desirable to have condensate trays – the DP of the primary air will be below room DP when a system starts up after overnight or week-end shutdowns. Like fan/coil units, air-throws are limited to approximately 5·5 m (18 ft), so that induction units are often used only for perimeter zones. Controls of these units can be complex and costly and, added to the costs of well insulated ductwork for the high-pressure primary air, this rarely enables the use of high pressure induction units to be cost-effective unless heating capabilities will be regularly required.

Variable air volume (V.A.V.) systems

With each of the equipment types previously described, the volume of air entering the conditioned space remains constant. Capacity control is effected by changing the capacity of the cooling coil, or by bypassing some air around it, thus changing the temperatures of the supply air. In the V.A.V. system, however, the amount of air delivered through the supply grilles is varied in proportion with the size of the cooling duty. Its temperature remains constant.

In its simplest form, the volume of air can be varied by face and bypass dampers in a terminal control unit. The dampers would be controlled by room temperature, and bypass an increasing quantity of supply air from the terminal unit to the return air duct, as space temperatures fall. Such units are located above suspended ceilings, the space above the ceiling acting as a return air plenum, or duct, into which unwanted supply air is 'dumped'.

Many variations can be played on this theme. More accurate control can be secured if electric or hot-water heating coils are installed in or close to the terminal units. And the use of return air ducts is no longer necessary if the air-conditioners which supply the conditioned air are designed to bypass unwanted supply air

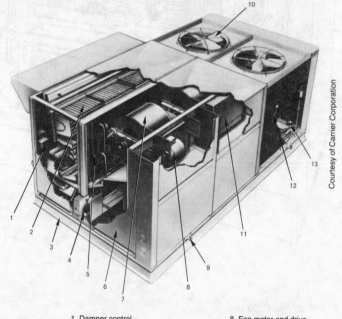

Fig. 7.28 V.A.V. terminal unit

Fig. 7.29 V.A.V. air-conditioner

1 Damper control
2 Filters
3 Roof curb
4 Static pressure control regulator
5 Hot gas bypass
6 Cabinet
7 Belt-driven evaporator fans
8 Fan motor and drive
9 Lifting lugs
10 Direct drive condenser fans
11 Control box
12 Compressors
13 Power and utility connections

COMMERCIAL AND INDUSTRIAL AIR-CONDITIONING EQUIPMENT

internally without its being delivered to and returned from terminal units some distance away. This is again done with the help of dampers, controlled by sensors reacting to the changes in the static pressure of air in the main supply duct. A self-contained air-conditioner, suitable for installation on a flat roof and designed especially for use with V.A.V. terminal units is illustrated in Fig. 7.29. Such units must, of course, be equipped with highly efficient capacity controls, to match low loads and the use of only small volumes of air. The use of multiple refrigeration circuits, the last of which is fitted with hot gas bypass, meets that requirement.

Figure 7.28 illustrates a cutaway terminal control unit. The damper position is controlled by the pressure exerted by a bellows, which expands – causing air volumes to decrease – as room temperatures fall; and vice versa.

One of the principal attractions of the V.A.V. system is that it enables the total volume of air supplied to a number of zones to be kept to the minimum needed to handle the maximum instantaneous heat gain. This is much less than the total of the volumes needed to satisfy the peak load in each individual zone; and enables savings to be made in the first costs and operating costs of fans and air distribution systems.

Air distribution

Supply air is either discharged direct into conditioned space by terminal units, or delivered through low, medium or high pressure ductwork. This is usually fabricated from galvanised steel sheet, with rectangular cross-sections for low and medium pressures, and circular spiral-wound design for high velocity installations. The use of thermal insulation is essential, or at least desirable, but careful design usually makes it unnecessary to use acoustic insulation in other than special cases – recording or television studios, for example, where installations must be virtually silent.

Fan types

Packaged airhandling units, and most fan powered terminal units, invariably use the forward-inclined centrifugal designs already described. We must however briefly examine axial-type fans. These do not discharge air at right angles to the direction of motion, like centrifugals; the airflow is straight through the fan blades, which are housed in a circular casing. The main components of typical axial fans, and the various names by which they may be known, as shown in Fig. 7.30.

This group of fans contains several 'families' of design. These range from simple propeller blade designs, to variable-pitch airfoil section blades in true 'axial flow' units, which have much in common with turbines. The difference in blade designs results in marked differences between performance characteristics. Propeller types can only operate at reasonable efficiencies when operating against low, or zero static pressures. Our most usual applications for them will be wall- or window-mounted exhaust fans for kitchens and toilets, or perhaps wall-mounted plant room ventilation units, where the plant room is on the perimeter of a building.

Axial fans are suitable for a wide range of pressures; single-stage models can handle the normal range of air-conditioning duties, which range up to about 250 mm (10 in) wg. It is also practical to achieve even higher pressures by multi-staging impellers – the fans then begin to resemble the multi-stage centrifugal refrigerant compressors described earlier in this chapter. Axials are, therefore, very versatile as well as efficient. The main factors inhibiting their use is size (in packaged airhandling equipment) and operating noise levels, which are higher than those of centrifugals.

Capacity control of axial fans can be

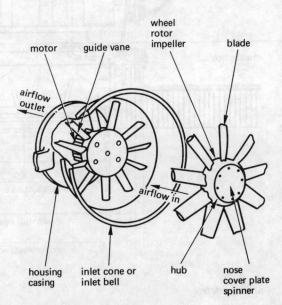

Fig. 7.30 Axial fan components

TROPICAL REFRIGERATION AND AIR-CONDITIONING

provided by either motor speed controllers or by using modulating pitch motor pulleys to vary the speed of fans themselves. A different approach has the blade pitch – the angle at which it bites into the air it is handling – variable through an internal mechanism.

Roof fans

Either centrifugal or axial fans can be used in roof-mounted, exhaust or air-intake fans. These are frequently used with ducting for such duties as exhausting air from toilets or internal plantrooms, or the positive supply of ventilation air to internal, ventilated areas.

Figure 7.31 illustrates typical examples of roof units for each class of duty. These units can be directly driven, or incorporate vee belt drives. Direct drive is perhaps the most desirable, since capacity control can then be provided by varying the motor speed. The use of motors mounted inside fan blades is increasing; it offers promise of much reduced dimensions, and simplicity of installation in ductwork.

Air registers and diffusers

A very wide range of options is available to the design engineer, for both supply or return/ventilation air duties. These are looked at in more detail, as they affect the installation and commissioning engineer, in Chapter 8.

Before ending this chapter, we should look briefly at some other classes of equipment which you may encounter, but are not in widespread use in tropical regions.

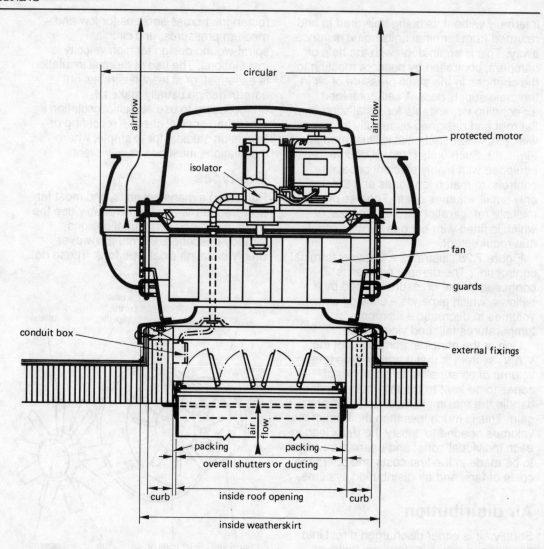

a) *Exhaust unit*

Fig. 7.31 Roof fan

COMMERCIAL AND INDUSTRIAL AIR-CONDITIONING EQUIPMENT

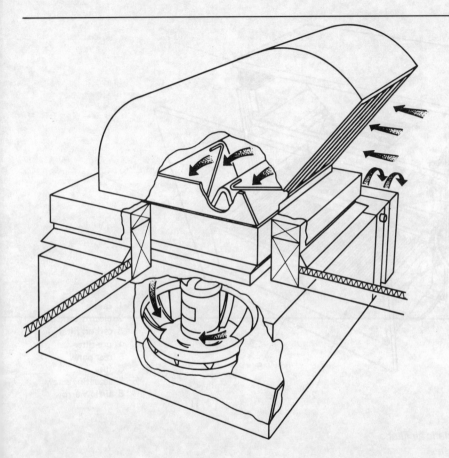

b) Intake unit

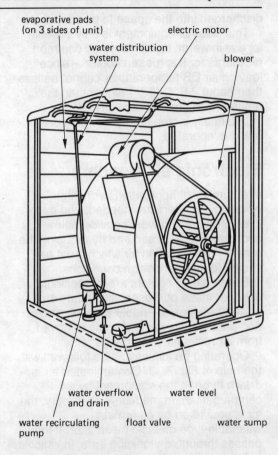

Fig. 7.32 Evaporative air cooler

Evaporative air coolers

The cooling effect secured by evaporating water or other liquids is the basis of refrigeration theory, and we need not elaborate on it. It is used also in cooling towers, or evaporative condensers, but we have not yet mentioned its application as a direct air-cooling method.

A typical air cooler is illustrated diagramatically in Fig. 7.32. A centrifugal fan draws warm, dry, outside air through pads of special wood or synthetic material which are saturated with water draining from overhead distribution troughs. On reaching the sump, water which has not been evaporated joins make-up water admitted by a float valve, and is pumped back to the distribution troughs. The water which does evaporate reduces the temperature of air passing through the filters, before it is

135

TROPICAL REFRIGERATION AND AIR-CONDITIONING

discharged into the space to be cooled.

This class of equipment is only suitable for use in warm, dry climates – a common nickname for it is 'desert cooler' – since leaving air DB temperatures cannot be less than about 2·8 °C (5 °F) higher than the ambient WB; and air humidity is substantially increased by the moisture being evaporated.

Electronic air cleaners

In a number of applications very high efficiency air filters are needed, and even in comfort-cooling work considerable economies can be secured by reducing the quantity of ventilation air which must be cooled to internal design conditions.

Electronic air coolers are increasingly used for these purposes, since they are efficient enough to enable tobacco smoke and odours, for example, to be removed from air recirculated through them.

Operating principles can be followed with the help of Fig. 7.33. Contaminated air is drawn through the equipment by a centrifugal fan. Immediately after entry, the air is prefiltered by a normal, washable-type metallic-mesh filter. It then passes through an ionising area, in which it is exposed to a high voltage, positive electrical field. This positively charges all solids in the airstream which now passes between a series of parallel collecting plates, which are alternatively positive and negative charged. The positively charged particles in the air are attracted to, and settle on, the negatively charged collectors. The air then passes through a charcoal filter, which removes any odours, before being returned to the room.

The collecting plates are cleaned of accumulated solids by removing and washing them. In addition to removing dust, this type of filter effectively removes bacteria as small as 2 to 3 microns, and is used to combat cross-infection in hospitals. It also has a number of industrial applications, including the removal and recovery of oil particles from mists or smoke generated in factories.

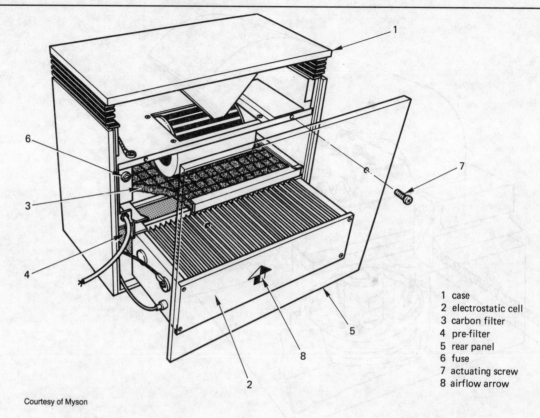

Courtesy of Myson

Fig. 7.33 Electrostatic air filter

1 case
2 electrostatic cell
3 carbon filter
4 pre-filter
5 rear panel
6 fuse
7 actuating screw
8 airflow arrow

Legionnaires' disease

In the 1970s, outbreaks of a mysterious illness with symptoms resembling those of pneumonia – which often proved fatal – affected hotel guests and also hospital

COMMERCIAL AND INDUSTRIAL AIR-CONDITIONING EQUIPMENT

patients in Europe and America. A mass outbreak in a hotel being used for a convention of ex-service men in America led to the illness being called 'Legionnaires' Disease' and triggered off an extensive investigation.

This isolated a bacteria, which was proved to be responsible. It appeared to flourish and multiply in warm water, and to be spread in fine droplets of water which people inhaled. However, no-one could be sure whether the organisms reached their victims directly from piped water supplies, or in water vapour originating from water cooling towers. They were in fact found in large quantities in some cooling tower sumps, and some newspapers, whose reports now appear to have been less than strictly precise, stated that the disease was spread by air-conditioning systems.

Later investigations proved that at least some incidents resulted from inhaling contaminated water vapour in shower baths, which are now being blamed for most, if not all, outbreaks. However, air-conditioning installations *may* have helped the bacteria to multiply, and to enter airstreams subsequently inhaled by victims. It therefore seems a sensible precaution to take steps to minimise or eradicate risks to people who use the installations for which we are responsible. The following precautions seem likely to be entirely successful:

a) do *not* locate fresh air intakes close to, or downstream (in terms of prevailing winds) of cooling towers, evaporative condensers, or other items of equipment which vent warm water vapour to atmosphere;

b) ensure that *all* drains are properly trapped, to prevent air blowing back through them into occupied areas;

c) thoroughly clean, and *sterilise with biocides* (i.e., chlorine or other strong antiseptics) all condensate trays in air handling units, and all surfaces in water cooling towers etc. *at least twice a year*. Where local conditions encourage the rapid growth of slime or algae, sterilise plant as often as required to prevent visible growth;

d) take all possible steps to prevent entrained moisture being carried out of equipment, into locations from which it might possibly enter occupied spaces.

These precautions cannot be guaranteed to remove all risks of future outbreaks; but current research results suggest that they are likely greatly to reduce possible health hazards.

Revision questions

1 An indirect R12 system is used to cool three large halls.
   a) Draw a diagram of the secondary circuit showing suitable controls for efficient operation.
   b) Suggest a suitable refrigerant control for the primary circuit.
   c) Suggest, giving two reasons, a good secondary refrigerant for this type of system.
   d) List two controls for the safety of the compressor and evaporator.
   (WAEC)

2 Draw a labelled diagram showing at least *six* main components of an absorption-type packaged water chiller.

3 Name, and state the functions of, two lubrication system controls which are usually fitted in large, centrifugal-type refrigerant compressors.

8. Pipeline and ductwork installation

Having examined typical system components we can get down to practical details of installation procedures for refrigerant and water pipework, and the erection and balancing of air distribution ducts and outlets.

Safety precautions

The syllabus which we are using properly places much importance on the use of procedures which are sound technically and in terms of the safety of both workmen and end users.

Safety warnings will be made whenever they seem appropriate. Before reviewing procedures in the workshop or on site, we suggest that you obtain, and use, safe clothing and protective equipment. The use of safety glasses and gloves have already been recommended. You could well add the following suggestions.

a) Do *not* wear ties or items of neck clothing or jewellery which hang outside your shirt. Around the world, ties become entangled in drive kits and strangle or break the necks of service personnel. Any metallic item which could accidentally complete a circuit between you, and live cables or contacts, should be worn only out of working hours.
b) Long sleeves should be buttoned down before you work with refrigerants, chemicals, brazing equipment, or gas or kerosene operated leak detectors.
c) Most of us will suffer an electric shock sooner or later, but you should try to avoid even one – they are dangerous. Those who have had theirs, and do not wish to repeat the experience, favour shoes with rubber soles, and rubber rather than canvas gloves.

Notes on contents

There are no short cuts along the road to experience, and Chapters 8, 9, 10 and 11 are intended only to complement the practical workshop instruction required by the syllabus and for which there is no substitute. We are not, therefore, attempting to cover every subject which may be scheduled, or taught in qualifying courses. Our objective is to summarise some of the more complicated and important procedures; and help students understand 'why' they are necessary, rather than exactly 'how' they are performed.

The successful diagnosis and rectification of operating faults often (when the plant concerned is not a self-contained, factory-assembled package) require the ability to recognise shortcomings in system layout or assembly, and we are including some basic design data to enable readers to check the logic of what might appear questionable performance characteristics. Developments of new materials and improved instruments lead to the constant introduction of more efficient procedures, and the adoption of revised performance standards. Like everything new, these changes are implemented more quickly in some countries than in others. Where it seems likely that there might be differences between procedures followed in some training centres, and the latest recommendations of respected professional bodies or specialist manufacturers, we will outline both; but we recommend the more effective of the two.

The importance of cleanliness

Modern systems use fast-running, precisely machined compressors with line fittings – such as refrigerant control valves – built to fine tolerances. System pressures and temperatures are high, and where hermetic compressors are used components are expected to work efficiently for anything up to 20 years without any internal maintenance.

That can only be done if the systems

themselves are scrupulously clean and free from all contaminants, when first assembled. The main contaminants which might affect systems are:
a) air, which in our trade always brings other contaminants with it;
b) moisture;
c) flux used when making joints;
d) metallic dust or swarf;
e) wax or other deposits from unsuitable oils.

The effects of contaminants on system components can be traced in the trouble-shooting data contained in Chapter 10, but some of the most obvious – and important – ones are noted now, for two reasons. It is essential that you take every precaution to avoid their entering systems on which you are working, and they often feature in examination paper questions!

The effects of contaminants

Air
In the unlikely event of any air entering systems being free from moisture, its effects will be limited to building up abnormally high pressures which might overload or cause major damage to the compressor motor. Since it cannot be condensed, dry air will accumulate in the condenser. This will become very hot as pressures build up, and compressor motors will overheat. Compressor operation will (with luck before any major damage results) be interrupted by the action of a thermal overload or – if fitted – the high pressure control. Unless corrected, excessive operating temperatures will cause the compressor lubricant to break down and to carbonise, or form *sludge*; any oxygen present reacts vigorously with oil.

Moisture
In classical refrigeration theory, the presence of moisture is signalled by the expansion valve of a low temperature system frosting over, as the moisture freezes and blocks the seat. In higher temperature, hermetically-sealed systems the results can be less visible but more damaging. Mixed with refrigeration oil and at high temperatures, water will lead to the formation of acids. These break down the oil and lead to sludge forming. They can also result in the compressor components being copper plated, or in the formation of metallic oxides. Motor failure or mechanical seizure can result. Motor failure can also follow if insulation is attacked by acids, and windings short to earth.

Flux
Soldering flux will, if allowed to enter a system, set up a number of chemical reactions with effects similar to those summarised under 'moisture'.

Metallic swarf
This may cause serious damage to compressor valves and/or valve plates, and pistons/cylinders and will block and/or prevent the correct operation of refrigerant valves and strainers. It is generally the result of careless cutting and preparation of tubing (see 'Cutting and Jointing Copper Tubing' on page 141).

Wax or other deposits
These come from the system lubricant, as the result of either moisture, or excessive operating temperatures, or the use of an incorrect type of lubricant. Use only the makes and grades of refrigeration oil recommended by the compressor manufacturer. Failure to do so may invalidate the guarantee as well as causing compressor failure.

Refrigerant pipelines

The wide differences in the chemical effects of fluorinated refrigerants such as R12, 22 and 502, and R717 (ammonia) on both metals making up refrigeration circuits and compressor lubricants, make it necessary to use separate system practices for each. In brief, 'Freon' circuit components are piped using refrigeration-quality copper tubing, and copper or brass line fittings; whereas R717 systems use heavy gauge steel equivalents. There are differences also in detailed practices used, for example, to deal with oil carried out of compressors by high pressure refrigerant vapour. Since the great majority of installations use R12, R22 or a member of their 'family' with similar characteristics, we will look in some detail at the best methods of assembling circuits for the fluorocarbons; and then note the essential variations made if working with R717.

Pipeline materials

Only refrigeration-quality copper tubing can safely be used in circuits with fluorinated refrigerants. This type has minimum wall thicknesses for all sizes (refrigeration tubing and fitting dimensions are quoted in

British/American unit outside diameters (OD). No metric dimensions are in common use at the time of writing). These are set out in Fig. 8.1.

OD Size	Soft Coils – Refrig. Duty	Hard Lengths – Type 'K'
$\frac{1}{4}''$	0·030"	–
$\frac{3}{8}''$	0·032"	0·035"
$\frac{1}{2}''$	0·032"	0·049"
$\frac{5}{8}''$	0·035"	0·049"
$\frac{3}{4}''$	0·035"	0·049"
$\frac{7}{8}''$	0·045"	0·065"
$1\frac{1}{8}''$	0·050"	0·065"
$1\frac{3}{8}''$	0·055"	0·065"
$1\frac{5}{8}''$	–	0·072"
$2\frac{1}{8}''$	–	0·083"
$2\frac{5}{8}''$	–	0·095"
$3\frac{1}{8}''$	–	0·109"
$4\frac{1}{8}''$	–	0·134"

Fig. 8.1 Copper tubing wall thicknesses

'Refrigeration quality' tubing is made to standards of annealing which suit it for bending and flaring; it is dehydrated and sealed before leaving the manufacturer's factory. The 'Type K' dimensions quoted in Fig. 8.1 are for straight lengths of hard tubing to A.S.T.M. standards. **It is not safe to use thin-wall plumbing tubing.** Refrigeration system lines must be capable of prolonged operation at internal pressures of 2 410–2 760 kPa (350–400 psi) and only purpose-made materials are acceptable.

Line fittings

The use of flared joints – made with forged brass frostproof flarenuts and unions, bends, tees, or elbows – is acceptable to sizes not exceeding $\frac{3}{4}''$, using SAE flare threads. System operating pressures make it increasingly difficult to secure absolutely tight joints using larger pressure type connectors. A full range of 'Solder-' or 'Wrot-type' fittings is available for tubing sizes from $\frac{1}{4}''$ to $4\frac{1}{8}''$ OD. These are manufactured from seamless copper tube, forged brass, or brass rod according to type. Unless there is every prospect that a fitting is likely to be removed from the system, the use of solder-type fittings for all $\frac{1}{2}''$ OD or larger joints is recommended.

Soldering and brazing materials

Loose use of the word 'solder' has apparently fostered the mistaken impression that joints can be made in soft solder. This is *not* the case. Soft solder, whether externally applied or in capillary-type fittings, is not suitable for use with any of the popular refrigerants; R22 in particular is liable to leak. This results from vibration as well as pressure and chemical action.

The use of silver solder, or less expensive silver solder substitutes, is recommended. A typical, high quality silver/phosphorous/copper alloy melts at a little under 650 °C (1 200 °F) and flows at 705 °C (1 300 °F). Another silver-bearing alloy contains silver, copper, zinc and cadmium. They require the use of a compatible flux, which becomes fully molten and active at or about the melting point of the silver solder itself, except where Sil-Fos is used in copper to copper joints.

In comparison with the above compositions and melting temperatures, a typical soft solder is made with tin and lead, and melts at about 230 °C (450 °F). It is good for electrical joints or low pressure work not exposed to consistent vibration.

The use of non-silver bearing 'brazing rod' is only recommended for joints which will *not* require to be opened and remade later. It operates at higher temperatures – typically, melting at 710 °C (1 310 °F) and flowing at 807 °C (1 485 °F). These are not so suitable for a number of line components, in particular solenoid valves, and the material is not as tough as the silver bearing alloys.

Soldering/welding equipment

Factory and workshop requirements are best met by electric welding equipment, which is available in compact designs suitable for site work. The main alternative is oxy-acetylene equipment, which is very effective, but usually inconveniently heavy and bulky for service other than major installation work. Less demanding needs to remove and replace line components or controls, or repair small leaks, are usually satisfied by compact LPG (Liquid Petroleum Gas) or LNG (Liquid Natural Gas) kits operated from replaceable cylinders of liquid gas.

Irrespective of the type of equipment used, it is essential that gloves and goggles are put on before torches are lit, or power switched on; and that flames are always directed away from the welder. Remember also that it is extremely dangerous to heat to welding temperatures tubes or components containing refrigerant or refrigeration oil. Before work is commenced, systems should be pumped out, and affected sections filled with inert gas – dry nitrogen is the best – or dry air. Do not use welding kit in spaces which are inadequately ventilated, or close to inflammable materials or potentially explosive chemicals such as refrigerant, or other high pressure gas cyclinders. Remove all insulation for several inches on either side of the area to be worked on; remember, if any fumes are generated, that those issuing from burning polyurethane are poisonous and that 'where there is smoke, there is fire'.

All equipment must be carefully handled and maintained, to make sure that jets are clean and in first class condition. Avoid knocking or jarring pressure gauges on oxy-acetylene cylinders, since they are sensitive and incorrect pressure readings can reduce safety as well as performance standards. Pay particular attention to instructions given in practical lessons, and watch the colours of flames and metals being joined as the jointing alloy begins to melt, and then to flow smoothly round the joint; in this instance, these visual indications of temperature are most useful.

Cutting and jointing copper tubing

Copper tubing seals should not be broken until the tubing is to be used, and the remainder of the coil should be resealed – using a pinching tool or a suitable plug – as quickly as possible. Avoid hacksaws when possible, and use a purpose-made tube cutter as shown in Fig. 8.2(a). Increase the blade depth in small steps. These cutters help to keep the cut absolutely square, and

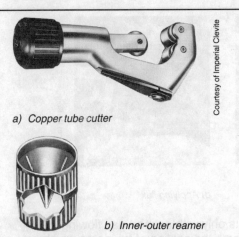

a) Copper tube cutter

b) Inner-outer reamer

Fig. 8.2 Tube cutting

Fig. 8.3 Lever-type tube bender

prevent the formation of burrs or swarf. Work with the cut end of the tubing held down, so that any swarf or dust does not slide into the pipeline; remove any burrs using either a reamer fitted to the tube cutter or, better still, an 'Inner-Outer' reamer as shown in Fig. 8.2(b). This cleans both the inner and the outer surfaces of the tube.

If tubing has to be formed into a bend before it is jointed, use one of the three types of aid available – external or internal bending springs, or a lever-type bender (Fig. 8.3). Each will help not only to make a better-looking bend, but to prevent tubing at the inside radius of the bend being distorted, collapsing and forming a restriction, or reacting to bending stresses by the formation of hairline cracks which will later be the sources of refrigerant leaks.

When the tubing is to be flared, *don't*

forget to slip on the nuts before it is too late! Lock the tubing in a proper flaring tool and lubricate the flaring cone with a little refrigeration oil. Many tools now have height indicators to show where the end of the tube should be located; if yours does not, experiment with the end of the tube flush with the top of the top of the bar for a start. Screw in the cone smoothly and steadily to the end of its travel. Unscrew and release cone and yoke, and make sure that the flare looks right, feels right and is free from cracks and distortions; then pull up the flare nut and check that everything fits perfectly.

When the joint is to be made with silver solder, polish the tubing with the emery paper whilst it is still held joint downward; and tap it to make sure that no swarf remains inside (Fig. 8.4). Slip it into the fitting to be used, and make sure that it fits and sits properly. Then with the end 6 mm

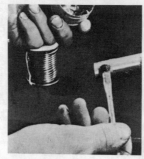

Fig. 8.4 Soldering   a) Burnishing copper tubing   b) Applying flux

($\frac{1}{4}''$) or so in the fitting, apply soldering flux (not needed with Sil-Fos on copper-to-copper joints) a little at a time, using a brush, to avoid getting any inside the tubing. Push the tubing home and rotate it to spread the flux as evenly as possible. Heat the tubing and fitting as evenly as possible, with the fitting joint pointing downwards and, as it warms up, touch the copper with the silver solder. As soon as this starts to melt, apply solder at say 90° intervals round the joint, and run the flame around the joint until solder flows, and is drawn into the joint and around the tube by capillary action. Make sure that the tubing is not overheated – if it is, a good and long-lasting joint is unlikely. Do not apply too much silver solder – just enough to produce a complete and visible ring at the top of the joint. This takes practice, and different techniques are needed when larger tubing sizes are silver-soldered. You'll learn them in the workshop, under skilled tuition.

Finally, remember our old enemy: contaminants. Make soldered or brazed joints only with an inert gas flowing gently through the section of tubing on which you are working. This will prevent the formation of oxide and scale inside joints. The nearer you can get to 100 per cent clean circuits, the more successful your installations will be!

Securing line fittings

Before fitting them, make sure that components such as solenoid valves and filter/driers are placed the right way round. Most manufacturers use an arrow to indicate the correct direction of flow where doubt can arise.

A number of line components can be damaged by excessive heat. Solenoid valves are obvious examples; but it is best to work on the assumption that *all* electro-mechanical items must be protected when being silver soldered into lines. Where possible, remove electrical components such as solenoid coils. Then protect the valve, or sight glass, or whatever else is in the way with an asbestos guard, or wrap it in fire-resistant material (the use of damp cotton waste does not fit in very well with our plea to keep systems clean, and free from moisture!) and *work with the flame pointed away from the body of the component.*

Evacuation and dehydration

When a system is fully assembled, but before pipelines are insulated, it is necessary to remove all moisture. This is done by drawing a vacuum throughout the circuit, so that any moisture will vapourise, and be removed by the vacuum pump.

This provides an example of long-established procedures not reaching higher standards which can be secured by the use of more recently developed equipment and methods, which are not yet in worldwide use.

Before reviewing alternative procedures, let us go back to basic theory. As we have noted, normal vacuum levels are measured and expressed in terms of mm – or inches – Hg below the pressure of a standard atmosphere. A perfect vacuum would equal approximately 762 mm (30 in) Hg; but normal gauges cease to read accurately at about 710 mm (28 in), which is also the limit which can be achieved by compressor-type vacuum pumps. High vacuums are expressed in terms of microns of remaining pressure, and measured with electronic 'high vacuum' gauges, which can operate down to about 20 microns. High vacuum pumps differ from normal 'compressor' designs.

A well established evacuation procedure is as follows: using a pump displacing 0·014 ltr/s (3 cfm) for systems up to 7·5 kW

PIPELINE AND DUCTWORK INSTALLATION

(10 hp), pull a vacuum of 710 mm (28 in) Hg *three* times adding R12 between each evacuation to bring the system pressure up to a gauge pressure of zero. After the third evacuation – this is known as the 'triple evacuation method' – the system is deemed to be free of all moisture, and ready to be charged.

The efficiency of the triple evacuation method is often 'proved' mathematically. The first evacuation will certainly remove about 90 per cent of all air and moisture vapour present. The second will remove 90 per cent of the remaining 10 per cent, leaving only 1 per cent, and 90 per cent of that will be extracted by the third evacuation.

These figures are mathematically impressive, but do not in fact offer proof that all liquid water present in the system will have been vapourised and removed. A vacuum of 710 mm (28 in) in an unheated system is not low enough to vapourise liquid water in short intervals between evacuation cycles. It is in fact highly probable that if there is any liquid water present in a system when triple evacuation is commenced, there will still be liquid water present when the third cycle is completed; and there is only one answer to the question, 'How much water is it safe to leave in a system?' – None!

The newer, and recommended, approach is the 'deep vacuum' method, in which the system is evacuated once only, to a vacuum of 100 microns or better. This requires the use of a high vacuum pump and a high vacuum gauge. It takes about an hour for a pump to pull a typical system down to 100 microns, and a hand valve

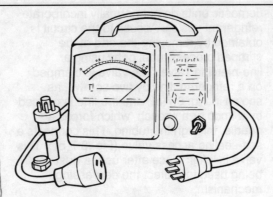

Fig. 8.5 *High vacuum gauge*

Fig. 8.6 *High vacuum manifold*

between pump and system should then be closed. If the gauge pressure does not rise, the system is free from moisture.

Referring to Fig. 8.5, high vacuum gauges should be calibrated over a wide range of pressures – 20 mm to 20 microns is preferred. They should register the total absolute pressure of all gases and vapour in the system.

High vacuum pumps should be used with a large bore manifold fitted with vacuum-tight valves, as shown schematically in Fig. 8.6. Note that connections are made to both high and low sides of systems. With valve P closed, the pump performance can be checked (and its oil changed if pressures are not satisfactory). With valve V closed and P open, the gauge shows the pressure within the circuit being tested.

Other leak test procedures

When systems using fluorinated refrigerants are too large for the use of high vacuum procedures, the triple evacuation method must be followed by a further leak test. Recommended procedures are as follows.

a) Add a partial charge of R12/22/502 as applicable, using gauges to ensure that system pressures do not exceed 689 kPa (100 psi).
b) Leak test all joints and components, using an electronic detector.
c) If no leaks are found, weigh in the balance of the system operating charge.

Notes
a) Any operating or safety controls liable to

TROPICAL REFRIGERATION AND AIR-CONDITIONING

be damaged by exposure to test pressures should be removed or blanked off.

b) Do *not* exceed the pressure quoted. There is a mistaken impression that test results are more conclusive if high pressures – perhaps 1 720 kPa (250 psig) – are used. Welded hermetic compressors in particular may be seriously damaged if exposed to internal pressures exceeding 1 034 kPa (150 psig). An electronic detector can be relied upon to pick up any leakage of halogenated refrigerants from systems at a gauge pressure of 350 kPa let alone twice that pressure.

Refrigerant charging

Refrigerant can be charged in liquid or gaseous form, using one of several types of equipment. As a general rule, liquid charging methods are used on new equipment requiring a known weight of refrigerant. The method is fast, and extremely accurate when charges do not exceed the limits (normally approximately 4·5 kg (10 lb)) of transparent, calibrated refrigerant vessels.

Gas charging is normally used to top-up systems. The procedure is slow, but enables gas volumes to be accurately controlled in accordance with the readings of suction and discharge gauges, an ammeter showing the power drawn by the compressor motor, and the absence of bubbles in liquid-line sight glasses.

Refrigerant control valves

Small welded hermetic compressors in domestic units do not normally incorporate refrigerant valves. Access to the circuit is obtained by removing the end of the crimped process tube, or by using a line-piercing valve. The valve is clamped on a refrigerant line by two screws, as shown in Fig. 8.7(a); and the line is pierced by a knob or a wrench, which forces a needle, through the tubing. This connects a self-sealing access valve (Fig. 8.7(b)). This valve is left in place after use, a dust cap being used to protect the depressing mechanism.

Intermediate capacity equipment with welded compressors normally incorporate access valves of the type shown in Fig. 8.7(b) in tails connected to high and low side pipelines, enabling refrigerant to be added at either the condensing unit or the airhandling unit, in the case of split systems. Good practice also suggests the use of a liquid-line valve having a port which can be used for the connection of a gauge or a capillary to control head pressure. In a number of cases, suction and discharge valves can be fitted (preferably by appliance manufacturers) to Rotolock connections on compressor casings (see Chapter 11).

Larger compressors, whether welded or accessible hermetic or open types, will have suction and discharge valves fitted as standard. The operating principles of these service valves and the liquid-line valves just mentioned, are the same; and can be followed from Fig. 8.8. The valve is operated by turning the control stem –

a) Assembly

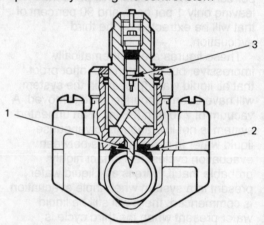

b) Detail

primary sealing
1 Needle valve shut-off with positive metal-to-metal seat provides leak-proof in-the-body seal
2 Rubber gasket forms a tight seal between tubing and valve

secondary sealing
3 Check valve seals flow passage

Fig. 8.7 Piercing valve

PIPELINE AND DUCTWORK INSTALLATION

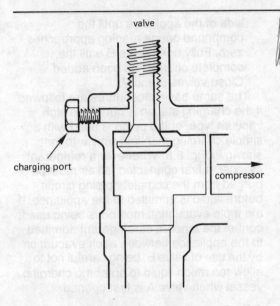

Fig. 8.8 Compressor service valve

appropriate lines. The gauge/access port is sealed, and gauges can be fitted or removed without loss of refrigerant.

Note
In some cases a front-seat port is also provided in suction and discharge valves, to enable safety or operating controls to be permanently connected. In this case, the controls can only be removed with the compressor switched off and the valve front-seated; but the suction or discharge line will be exposed to atmosphere whilst work proceeds.

Cylinder positions
The logical positions of refrigerant cylinders are:
a) For liquid charging – cylinder inverted above the system, charging valve down.
b) For gas charging – cylinder vertical, valve uppermost, beneath the system.

Charging equipment options

Conventional charging hoses
These are used to connect main items of equipment. For use with fluorinated refrigerants, they are normally in heavy duty neoprene tubing rated for use to 5 170 kPa (750 psi). Both ends have $\frac{1}{4}''$ SAE swivelling connectors, one of which is angled and incorporates a valve core depressor to automatically open (when tightened) or close (when slackened off) self-sealing access valves.

High vacuum hoses
These are of flexible, seamless metal construction; normally $\frac{3}{8}''$ inside diameter and 2 m (6 ft) long.

using a ratchet spanner – clockwise to 'front-seat' the valve, or anti-clockwise to 'back-seat' it. There is an intermediate, 'cracked' position. The purpose of these three positions is listed below.

Front Seated The valve is fully closed, sealing the compressor from the suction or discharge line. The charging port plug cannot be removed without loss of refrigerant.

Cracked All valve ports are open. A gauge, if fitted, shows the system suction or discharge pressure.

Back Seated This is the normal operating position. The compressor suction/discharge ports are open to the

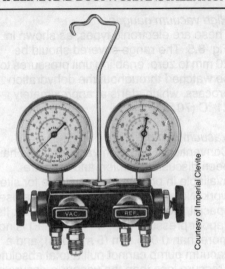

Fig. 8.9 Service manifold

Service manifolds
These are arranged for one suction and one discharge gauge; and either three or four flared connectors for lines from the vacuum pump, refrigerant cylinder, and the appliance to be tested. High vacuum designs have special, diaphragm-type soft-seat shut-off valves; a $\frac{3}{8}$ in vacuum port; and oversize internal passages.

Conventional gauges
Conventional gauges are normally 60 mm ($2\frac{1}{2}''$) in diameter, with $\frac{1}{8}''$ pipe thread connectors. Normal calibration ranges are 0–3 400 kPa (0–500 psi) for pressure gauges, and 760 mm Hg – 800 kPa (30 in Hg – 120 psi) for compound types.

High vacuum gauges

These are electronic types, as shown in Fig. 8.5. The range covered should be 20 mm to zero, enabling unit pressures to be watched throughout the dehydration process, which starts at approximately 21 °C (70 °F).

Vacuum pumps

Conventional and high vacuum types have been discussed earlier, and both are available in portable sets suitable for site work. Note that high vacuum models use a special, high quality paraffin-based oil. Its vapour pressure at 37·7 °C (100 °F) is not more than 0·005 mm (5 microns); and a vacuum pump cannot pull a total absolute pressure less than the vapour pressure of its sealing oil.

Charging stations

Many models, portable and immovable, are available. They normally incorporate a vacuum pump; transparent, calibrated refrigerant vessels for each important fluorinated refrigerant; gauges; valves; piped circuits; and connections to complete. A typical but simplified layout schematic is shown in Fig. 8.10.

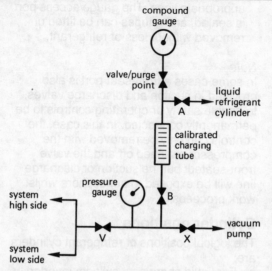

Fig. 8.10 Charging station piping

Charging procedures

The following sequence would be followed if using a charging station piped as in Fig. 8.10.

a) Close all valves.
b) Connect the appliance, preferably through a yoke connecting the charging panel to both high and low sides.
c) Open valve X. Switch on the high vacuum pump and confirm that it pulls down quickly to a high vacuum. If it does not, change the pump oil.
d) Open valve V. Watch the high vacuum gauge to confirm that the pointer falls steadily into high vacuum (this is not possible if there is a substantial leak in the system).
e) When the gauge reads 100 microns, close valve X and switch off the vacuum pump. If the vacuum gauge reading remains steady, close valve V.
f) Open valve A to admit the required volume of liquid refrigerant to the calibrated charging vessel. Close valve A when the correct level is reached.
g) Crack valve B and allow refrigerant to flow slowly into *only* the high pressure side of the appliance until the compound gauge reading approaches zero. Fully open valve B until the complete charge has been added. Close valves B and V.

The same basic procedures are followed if the charging station is not of the high vacuum type, or if a pump is used with a simple charging manifold similar to that shown in Fig. 8.6. Whenever a refrigerant cylinder is first connected, all air must be purged from the complete piping circuit before liquid is admitted to the appliance. If the triple evacuation method is being used, control the amount of refrigerant admitted to the appliance between each evacuation by the use of valve B, being careful not to allow too much liquid to enter the charging vessel when valve A is first opened.

Oil charges

It is usual for an operating charge of oil to be provided by compressor manufacturers; but it is always desirable to seek confirmation from the compressor *supplier* – whether an 'OEM' (Outside Equipment Manufacturer) appliance manufacturer or a compressor wholesaler – that replacement compressors *do* hold a full charge of lubricant.

Oil is added to welded hermetic systems, on the rare occasions when a top-up charge is needed, through the compressor service-tube or a high side line-piercing valve. Some charging stations include an oil metering cylinder piped into the circuit so that it can be used in the same way as the refrigerant vessel. When compressor casings are not of welded hermetic design, an oil charging plug is built in to the crankcase. Oil is then added with a hand pump, as described in Chapter 11.

Freon piping practices

It is neither possible nor necessary to provide a treatise covering all aspects of refrigerant line sizing and layout. When working with contracting companies, you can expect to be given at least a schematic drawing showing the positions of all system components, and all line and connection sizes.

As background to the use of such contract drawings, and to help when installing simple equipment – such as domestic split systems – which do not warrant installation drawings, the more important basic requirements are briefly outlined in the next few paragraphs with illustrations.

Oil return and refrigerant velocity

As we have seen, when fluorinated refrigerants are used – and particularly in the case of R22 – oil constantly circulates around the system, moving much more easily in liquid than in gaseous refrigerant. This is true only of properly designed circuits; if refrigerant gas velocity is too low, oil will separate out and collect in any convenient traps or pockets. At the best, operating performance will be much reduced; at the worst, compressors will seize or their motors will burn out because of oil shortage in the crankcase.

Full details of requirements, and tables of the pressure drops caused by tubing and line fittings at various suction and condensing temperatures, are contained in publications such as the ASHRAE Handbook of Fundamentals, and manuals prepared by leading equipment manufacturers. These provide tables of recommended sizes of liquid, suction and discharge lines for each of the main refrigerants, over a full range of operating conditions.

Examples of the minimum and maximum gas velocities of R22 required in air-conditioning installations in order to ensure good oil return characteristics and avoid noise or vibration are:

minimum velocity in horizontal lines = 2·5 m/s (500 ft/min)
    in vertical risers = 5·0 m/s (1 000 ft/min)
maximum velocity in any line = 20·3 m/s (4 000 ft/min)

Refrigerant lines are sized, with reference to the volumes of liquid and gas needed to meet performance needs, to ensure that total pressure drops do not exceed (again for R22 air conditioning work):

hot gas lines = 41 kPa (6 psi)
suction lines = 21 kPa (3 psi)
liquid lines = 21 kPa (3 psi)

Pressure drop through liquid-line risers is critical, since if the pressure of the liquid is reduced too much, it will 'flash' into pockets of gas. This upsets expansion valve operation, and system capacity. The only way the problem can be avoided is to subcool the liquid refrigerant after it has condensed. Obviously, there are limits to what can be done and riser heights in excess of 10 m (30 ft) require particular attention at the design stage.

Oil traps

Traps are extremely simple, but do two things: First, a trap provides a container into which any free oil can drain, and accumulate. Second, as the oil level rises the refrigerant velocity is automatically increased – it has to pass through a smaller area. Increased velocity increases the oil-carrying capacity of the gas, and oil is entrained and lifted up the riser.

An oil trap is needed at the bottom of any suction or hot-gas riser exceeding 2·5 m (8 ft). Hot gas risers over 9 m (30 ft) high should be fitted with oil separators rather than oil traps (see Fig. 6.10). Figure 8.12 illustrates recommended piping practices for evaporators installed above and beneath compressors.

In the case of Fig. 8.12(b) the riser loop prevents any liquid from draining into

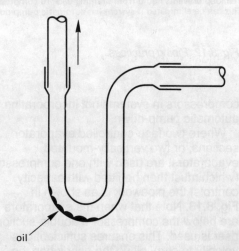

Fig. 8.11 Oil trap section

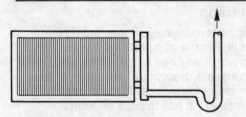

a) Compressor above evaporator
The trap ensures that liquid refrigerant and oil drain away from expansion valve phial. Make trap as short as possible to minimise the amount of oil.

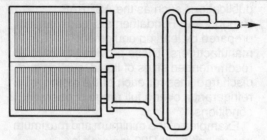

a) Multi-section evaporator, compressor above
Fit double pipe riser if necessary.

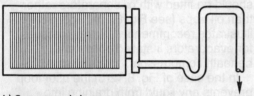

b) Compressor below evaporator
The loop prevents liquid from draining back to compressor. This can be eliminated if system has automatic pump-down.

Fig. 8.12 Piping practices

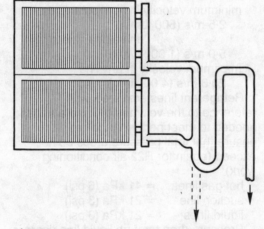

b) Multi-section evaporator, compressor below
Flow from upper evaporator can not affect valve phial of lower evaporator.

Fig. 8.13 Piping practices

compressors in systems not incorporating automatic pump-down.

Where two face-controlled evaporator sections, or two vertically-mounted evaporators, are used with one compressor (which must then be fitted with capacity controls) the pipework is as shown in Fig. 8.13. Note that when the evaporators are below the compressor a double-suction riser is used. This ensures sufficient gas velocity to carry oil up the riser when the compressor is unloaded.

Pitching horizontal gas lines

To further help any free oil to return to the compressor crankcase, 'horizontal' runs of hot-gas or suction lines should be pitched downwards slightly – 12 mm per 2 500 mm (½ in per 10 ft) is enough – *in the direction of gas flow*.

Line valves

It is advisable to install manually operated valves to isolate those components – such as strainer/driers – of field-assembled plant, which may need to be removed or replaced. In larger installations, valves enabling remote condensers and evaporators to be isolated can secure major savings in the event of physical damage to those components or refrigerant leaks.

Flexible connections

Anti-vibration hoses should be fitted in suction and discharge lines to prevent the transmission of noise and vibration through building structures. One must select longer rather than shorter hoses since they can then satisfactorily take up pipeline thermal expansion and contraction without being stretched.

Pipe hangers/supports

Pipe hangers or supports should be spaced not more than 2 to 3 m (6 to 10 ft) apart with closest spacing for smaller sizes, and not more than 0·6 m (2 ft) from each bend. Hangers should fit *over* any pipeline insulation, or over circular timber or plastic 'crocodile' spacers of the same thickness as the insulation. Insulation should be protected by galvanised steel sleeves (which are also necessary when pipes pass through masonry). The use of threaded rod type hangers which enable heights to be

PIPELINE AND DUCTWORK INSTALLATION

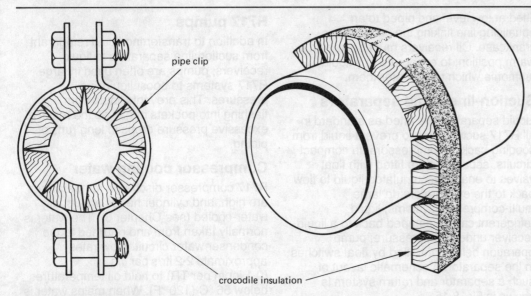

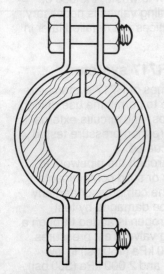

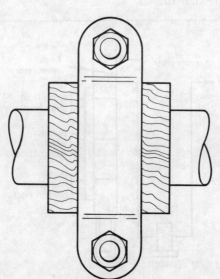

a) Crocodile insulation

b) Hardwood rings

Fig. 8.14 Pipe supports

accurately matched, is usual. When pipe runs are long, the use of hangers designed to permit pipe expansion or contraction is necessary.

Pipe insulation

The use of foamed insulation with inherent vapour seals is usual. The thickness required to prevent the formation of condensation varies with pipeline temperature and space dew point – obviously, a suction line in a low temperature coldstore requires insulation with a higher 'R' value than that used on an air-conditioning system. Remember that *all* joints must be vapour-sealed, and that insulation is best secured and sealed by a suitable adhesive.

R717 system practices

There is a dangerous tendency to imagine that the steel refrigerant lines used in ammonia installations need not be so thoroughly cleaned or dried as fluorocarbon systems. This is not the case – R717 is a powerful solvent, which will scour pipe and fittings of any dirt, scale, sand or rust which might be present; and these are as damaging to compressors used with R717 as to any others. The use of liquid-line filter/strainers, and suction port strainers in compressors, is essential.

The heavyweight steel tubing used for lines together with purpose-made, weld-type steel fittings, must be rated for use at pressures of at least 300 psi. Joints not larger than 32 mm ($1\frac{1}{4}$ in) can be threaded provided fittings are of 2 000 psi, and a joint sealant such as litharge is used.

Joints to components such as strainers and float controls, which must be opened for maintenance, are made with weld-type flanges. These provide a welded joint to the pipe, and a flanged connection to the control or fitting. To ensure that all flange dimensions are fully compatible, it is best to purchase flanged fittings with matching counter-flanges provided by the equipment manufacturer. Pipe hanger spacings should be 2·5–3 m (8–10 ft). Note that all gaskets must be made from non-metallic, fibrous material: and all containers which can be valved off *must* be fitted with automatic pressure relief valves.

System oil requirements

An oil separator must be fitted in the discharge line of each compressor in an installation. It should be as far from the discharge port as possible, to enable the refrigerant to cool down before entering the separator. Construction is basically similar to that shown in Fig. 6.11, but special precautions are taken to prevent any liquid refrigerant from flowing back to the crankcase when the float valve opens.

As oil separators are not 100 per cent efficient, and oil is not readily miscible with R717, any remaining lubricant will accumulate in system components. Oil is more dense than R717, and will form a separate layer beneath liquid refrigerant. This is particularly undesirable in the evaporator, where surfaces would be insulated and heat exchange capacity reduced. In practice, the use of oil drainage lines is common in large systems. These are often connected into an oil receiver fitted at low level and piped to an equalising line linking the compressor crankcase. Oil receivers must be kept in a warm position, to evaporate any liquid ammonia which might enter them.

Suction-line liquid separators

Liquid separators are fitted as standard in all R717 suction lines to prevent liquid from flooding back to compressors. In compact circuits, separators are fitted with float valves to enable accumulated liquid to flow back to the evaporators. In large multi-compressor systems, liquid refrigerant can be pumped back to a liquid receiver under high pressure, pump operation being controlled by float switches in the separator. A schematic layout of such a separator and return system is shown in Fig. 8.15.

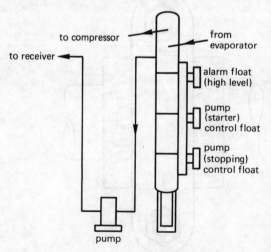

Fig. 8.15 R717 suction trap and liquid pump

R717 pumps

In addition to transferring liquid refrigerant from suction line separators to liquid receivers, pumps are often used in large R717 systems to boost liquid line pressures. This prevents liquid from flashing into pockets of gas as the result of excessive pressure drops in long runs of piping.

Compressor cooling water

R717 compressor discharge temperatures are high, and cylinder heads are usually water-cooled (see Chapter 6). The water is normally taken from and returned to the condenser water circuit. Flow rates approximate 2·2 ltr/s per J/s (0·1 igpm per TR) to hold oil temperatures below 56 °C (120 °F). When mains water is used, and then run to waste, the use of thermostatic throttling valves is necessary. Water piping practices are detailed later in this chapter.

Leak testing R717 systems

The internal volumes of R717 systems are normally too large to permit the use of high vacuum techniques, and circuits external to the compressor are then pressure tested before being evacuated.

After blowing through the pipework to remove any scale or solid contaminants and blanking off the compressor and any controls liable to be damaged by test conditions, dry nitrogen is added through a pressure-reducing valve. Test pressures favoured are 1 000 kPa (150 psi) in low side components, and 2 000 kPa (300 psi) in the high pressure side. The ability of the system to maintain test pressures for several hours must be confirmed, using accurate gauges. If leaks are found in welded joints, they cannot be cured by additional welding rod; the joint should be cut out and remade.

After dry nitrogen is discharged to the atmosphere – hopefully carrying with it any water vapour which was present in the circuit – the system is evacuated to the best vacuum which can be secured. In this connection, any vacuum less than 710 mm (28 ins) Hg will *not* vapourise liquid water, unless the system temperature is above 21 °C (70 °F).

After being evacuated, circuits are partially charged and leak tested before the remainder of the R717 is added through a charging valve in the high pressure side. One's nose should be as good a detector of leaks of R717 as the devices described in Chapter 4.

Water piping

Many of the installation practices recommended for refrigerant lines apply to condenser and chilled-water circuits. This is true of the need to use only clean tubing, and purpose-made bends, tees, elbows etc. and applies also to the use of anti-vibration couplings, whether of hose or spherical design, and to pipe hangers/supports. The inclusion of supports and couplings which can take up expansion is most important in the case of chilled-water circuits; and where long horizontal runs are in trenches, the use of roller chair assemblies is desirable. Catalogues issued by leading manufacturers of piping accessories contain such data as the weight of pipes and water contents, recommended hanger spacings and the weights which their own products can safely support. Any ferrous (iron-containing) lines run underground, other than in dry and ventilated pipe trenches, must be protected against corrosion (which could also result from exposure to industrially contaminated air) by suitable wraps or paints, where no insulation is used.

Chilled-water circuits

Pipe sizes
Water velocities are normally within the range 0·6–2·4 m/s (2–8 ft/s) and cause pressure drops of 0·6–3 m (2–10 ft) of water per 30 m (100 ft) of *effective* pipe length. Effective length is the sum of actual length plus the resistances offered by such system components as bends, tees, and – in particular – valves. Component pressure drops are summarised in detailed tables; but for present purposes, a rule of thumb that effective length will be about twice the actual length is all that you need.

Water volumes depend on the temperature difference between water entering and leaving the chiller. Typical temperature ranges, and resultant water volumes, are summarised in Fig. 8.16.

Pipe material
As chilled-water lines are closed circuits and externally insulated, and their water content chemically treated to inhibit scale formation, medium-weight black steel tubing is normally used. Joints up to and including 65 mm (2½ in) outside diameter can be screwed, joints being sealed with any material suitable for pressures of at least 345 kPa (50 psi); and all larger size connections are welded. Weld-type flanges should be used for controls etc. Accessories used must include water make-up/expansion tanks (low pressure or open type if installed above the highest point of the system, or pressurised if at low level), drainage valves at low points and manually or automatically operated air vents at individual high points. Typical connections to cooling coils are illustrated in Fig. 8.17.

Temperature difference		Water volumes	
°C	°F	m^3/h per 1 000 kCal/h	igpm per 1 TR
4·4	8	0·226	2·5
5·6	10	0·181	2·0
6·7	12	0·151	1·67
7·8	14	0·129	1·43

Fig. 8.16 Chilled-water volume: temperature relationship

TROPICAL REFRIGERATION AND AIR-CONDITIONING

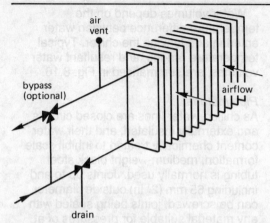

Fig. 8.17 Water to cooling coil connections

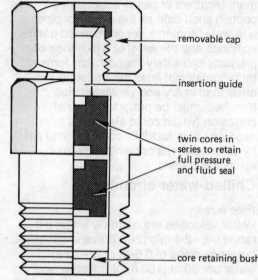

Fig. 8.18 Test plug section

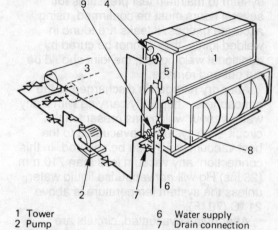

1 Tower
2 Pump
3 Condenser
4 Water gauge
5 Bleed-off line
6 Water supply
7 Drain connection
8 Overflow connection
9 Filter

Fig. 8.19 Typical condenser water circuit layout

Insulation
Remarks on refrigerant suction lines apply to chilled water circuits. Pipes passing through masonry must have sleeves of galvanised steel or another metal.

Condenser water circuits

Pipe sizes
Water velocity and pressure drop ranges are as for chilled-water circuits.

Water volumes depend upon the temperature rise through the condenser. When using recirculated, cooling-tower water the temperature drop is approximately 4·4 °C (10 °F), which gives water volumes of 0·2 ltr/s per 3·5 kJ/s (2·5 igpm, or 3 usgpm, per TR).

Pipe material
Medium-weight, galvanised steel tubing is used for the majority of condenser water circuits, with jointing methods as in chilled-water systems. It is obviously essential that any damage to galvanising resulting from welding must be offset by the generous use of zinc or other rustproofing paint on both internal and external surfaces.

Instrumentation
Adequate provision must be made for the use of flow meters, thermometers, and pressure gauges necessary to balance water flows through new circuits or to test performances of existing plant. Unless the instruments are to be left permanently in place, the most cost-effective approach is the installation of self-sealing, universal test plugs to receive instruments fitted with compatible probes. No isolating valves are necessary.

A typical condenser water piping schematic is shown in Fig. 8.19. Note the bypass line with volume control/shut-off valve, which can be used to finely balance water volumes through the condenser.

Water treatment
Scale inhibiting chemicals for chilled-water circuits are normally inserted when the system is charged, and need be checked only once a year, unless there has been a major leak.

Chemicals for condenser water systems are usually 'drip fed'. They are dispensed in cooling towers by perforated plastic

PIPELINE AND DUCTWORK INSTALLATION

containers, or from tubes through which part of the total water flow is diverted. These need to be checked monthly. Kept at correct strength, they can save the needless hard work required to remove accumulated scale from condensers.

Drainage lines

Branch drainage lines, and often risers also, can be run in lightweight copper tubing with capillary type, soft solder fittings, or plastic tubing with chemically 'welded' joints. In the case of high rise buildings or other projects in which it might be necessary to drain down chilled or other water systems operating under high static pressures, materials must be selected with reference to the highest pressures to which they might be exposed under extreme circumstances.

It is particularly important to remember that the drainage lines from the condensate trays of airhandling units should *not* be interconnected, before entering the main drainage riser, with drainage connections in chilled or condenser water circuits. This practice incurs the risk of water under considerable pressure flooding back into, and overflowing from, condensate trays.

The use of properly designed and connected traps is most important, and we should not lose sight of the fact that condensate often leaves airhandling units at temperatures requiring the use of thermal insulation on drainage lines.

Drainage sump tanks, with float-switch controlled pumps, are required when water-carrying equipment is housed in plantrooms below the level of mains drainage systems.

Water pumps

Centrifugal pumps are invariably used for water services, single-stage construction being suitable for most applications which we are likely to encounter. As with compressors, there is a choice of manufacturing methods – independent pumps either directly driven, or fitted with vee belt drives; and monobloc construction units, in which the pump is in the housing that contains the motor, both sharing a common shaft. A second variable is operating speed, 1 450 and 2 900 r.p.m. being the normal 50 Hz options.

The most logical choice is that of a monobloc type, since the construction eliminates both vee belts – which are always subject to wear – and any need to watch pump and motor alignment. Poor alignment invariably results in damage to both the drive couplings and the seal of the pump shaft; it is a risk that is best avoided. This leaves the question of speed. The higher speed provides higher head characteristics, but often causes objectionable noise levels; so that when possible 1 450 r.p.m. (or 1 750 on 60 Hz supplies) is the popular choice.

Regardless of type, or speed, it is essential that all pumps and driving motors are securely bolted to adequate – and preferably well isolated – baseplates or foundations. Pumps must also be isolated from pipelines. The fittings shown in Fig. 8.20 are recommended in condenser circuits.

Control of water volumes and system head pressures is more critical in chilled than condenser water circuits; and more complex circuit layouts – often having a number of separate pumps and supply and return lines piped into a ring mains – sometimes necessitate the use of non-return valves. A typical fittings schematic is therefore as shown in Fig. 8.21.

Insulation and hanger/support

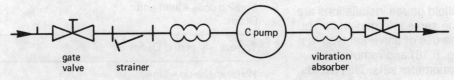

Fig. 8.20 Condenser pump fittings

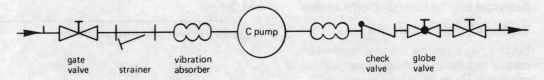

Fig. 8.21 Chilled-water pump fittings

TROPICAL REFRIGERATION AND AIR-CONDITIONING

arrangements for piping have already been noted. Pipeline valves etc. can be heavy, and require support by floor-mounted guides or anchors. When needed, these are often fabricated on site to secure an exact fit.

Water circuit instrumentation

Provision must be made for either the permanent installation, or the connection and removal as and when required, of all gauges, flow meters, and thermometers necessary to balance and accurately establish the performance of all main chilled and condenser water circuit components.

In the case of pumps, a manifold gauge arranged as in Fig. 8.22 can be fitted. The layout enables us to establish the pressure differential across the vital operating components, without distortions which might result from the presence of scale or other foreign matter in strainers. A more comprehensive summary of recommended instrumentation provisions appears as Fig. 8.23.

Whilst manifold gauge installations are permanent, the same information can be secured by the use of universal-type test plugs (see Fig. 8.18) and removable differential manometer sets. This avoids the possibility of gauges being damaged, and giving false readings; and the test plugs can also be used for thermometers.

Water circuit balancing

Since it has a narrow thermal band width, careful regulation of chilled water volume is essential. There must be a close relationship between design and actual

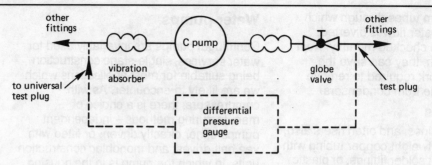

Fig. 8.22 Manifold gauge arrangement

Location	Differential pressure gauge	Pressure gauge	Thermometer	Test plug
Pump – Inlet and Outlet	x	x		x
Condensers – Inlet and Outlet	x		x	x
Chillers – Inlet and Outlet	x		x	x
Cooling coils – Inlet and Outlet			x	x
Towers – Inlet and Outlet				x
Water make-up – Line				x
Zone water mains – Inlet and Outlet	x	x		x

Fig. 8.23 Water system instrumentation

levels of flowrate, pressure drop, and temperature change through evaporators and condensers. This is most easily established when all equipment is new, free from wear (and scaling!), and operating in close conformity with manufacturers' performance data. We recommend that you work to a consistent programme when balancing water circuits.

a) Using pumps only, and after having purged the system of all air or other contaminants, balance water flow rates as shown in design details, using flow meters.
b) Check pressure drops through individual heat exchangers, and through the system as a whole (using gauges).
c) After starting the refrigeration system and allowing water temperatures into and out of evaporator and condenser to approach design levels, finely balance water flow rates to achieve design temperature differences under full load conditions. Note and record *all* final settings and readings.

Air distribution

Ductwork

Glass fibre ducting – which is self-insulating – is often used in domestic and other small installations where air velocities do not exceed 10 m/s (2 000 ft/min) and duct dimensions are not so large as to require structural supports. In contrast, the majority of commercial and industrial supply and return air ducts are fabricated in galvanised steel or aluminium sheet. In most cases rectangular ducts are used in low pressure (up to 50 mm (2 in) wg) and medium pressure (50 to 125 mm (2 to 5 in) wg) jobs; and spiral wound, circular construction for high pressure (100 to 250 mm (4 to 10 in) wg) installations. Duct sizes are usually designed to ensure relatively consistent friction losses throughout the system; and good design practices minimise air turbulence, which increases both pressure drops and noise levels. Flexible ducting is often used to connect ducts and diffusers: this must not be needlessly long, and must be insulated.

It is also essential that air leakage from ducts is kept to a minimum – a loss of 10 per cent by volume can easily result in a temperature increase of 1 to 1·5 °C (2 to 3 °F) in the conditioned space. Construction standards covering sheet metal gauges, jointing methods, and 'cross breaking' or other transverse reinforcements to prevent 'drumming' are defined in detail by authorities such as SMACNA/ASHRAE, and HVCA/CIBS. If these standards are combined with good design practices, systems will prove easy to balance and it will not be necessary to use acoustic insulation on other than very low noise level applications such as broadcasting or recording studios. The use of external thermal insulation will however be essential in most cases, to beat our old enemy: condensation. The insulation can take the form of foamed plastic or mineral fibre boards or sheets, the only essentials being that it is thick enough and that installation procedures ensure that it is effectively and permanently vapour-sealed.

The installation team has no responsibility for design matters, but is well advised to make sure that systems include:
a) flexible connections between airhandlers and ducts, to prevent transmission of vibration or noise;
b) robust and easily adjusted duct hangers. These normally have threaded rod suspension with angle iron cross-straps at a consistent height;
c) sealed seams and joints, to minimise air losses;
d) layouts which avoid radical changes of direction within short distances of each other; or fittings which change more than one rectangular duct dimension;
e) airfoil-shaped air-turning vanes in all bends or elbows;
f) fire dampers to isolate plantrooms from conditioned space and to prevent the rapid spread of fire from floor to floor, or from zone to zone;
g) volume dampers which can be securely locked at or close to the commencement of each branch duct;
h) seals over all openings in ducts and airhandling equipment, to prevent the entry of dust, dirt etc. before the system is ready to be balanced.

Air grilles and diffusers

A wide range of horizontal-discharge supply air outlets for use with underceiling duct runs, and diffusers for use on ceilings or overhead ductwork, is available. These are sized to achieve a good balance between: air volume, length of throw and air jet velocity.

Errors generally result in draughts or noise. In general terms, ceiling diffusers are preferable to horizontal-discharge registers: but in this section we are more concerned to see that either type is fitted with adequate volume and directional controls.

When registers or diffusers are installed in stub ducts which are either horizontal or vertical to the main airstream, our first requirement is for air equalising vanes as shown in Fig. 8.26. These do not govern the volume of air admitted to the stub, but equalise the airstream over the entire area of the outlet. Next, we need volume control dampers, which will be of different design for horizontal-discharge and ceiling outlets. There are a number of options within each category. For horizontal-discharge grilles we recommend the use of opposed-blade dampers with each blade pivoting at its centre. Each blade moves towards its neighbour when closing; and the design produces a number of small airstreams which quickly expand and cover the full face area of the grille. This type of damper – also extensively used as an in-duct volume controller – is set by a locking key, which is accessible through the blades of registers when forming part of a volume-controlled outlet.

There are also many options for volume control in ceiling diffusers, the multi-louvre damper illustrated in Fig. 8.27 being most popular. Like the opposed-blade damper, it splits the airstream into a number of small 'slices' which quickly expand to fill the entire outlet area.

Most side-throw registers are available with two blades which enable the angle of airstream to be adjusted in the vertical plane, by varying the inclination of the horizontal blades, which are normally at the rear. The front, vertical blades are similarly

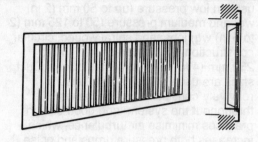

Fig. 8.24 Horizontal-discharge register

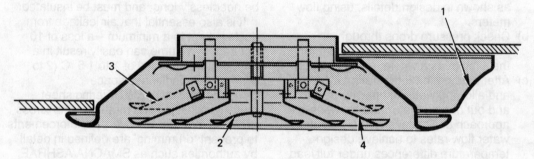

Fig. 8.25 Ceiling diffuser section

1 Anti-smudge ring
2 Centre cone regulator
3 Cones in raised position vertical discharge
4 Cones in lowered position horizontal discharge

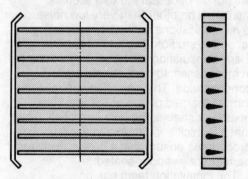

Fig. 8.26 Air equalising vanes

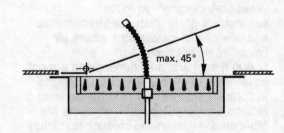

Fig. 8.27 Multi-louvre damper

adjustable to vary the width of the airstream in the horizontal plane. The wider the setting of the vertical blades, the shorter the distance which the air will carry. Throw can however be increased by angling the rear, horizontal damper so that the airstream leaves the register about 15° above the horizontal.

In the case of ceiling-mounted diffusers, the intention is that incoming air should run close to the ceiling for some distance, inducing room air, before falling into the occupied zone. In some cases the distance between horizontal cones or plates can be adjusted if necessary, enabling the air pattern to be changed to provide a finer balance.

In some jobs – particularly where air volumes are large – perforated ceiling tile type supply diffusers are used, to secure reduced velocities. In extreme cases, such as *white rooms* where air velocities must be kept to the minimum possible, a whole wall can be used as an air supply area; with the opposite wall forming the return grille.

Air velocities

Air velocities are, as stated, related to noise levels. The range of velocities most used in wall-mounted applications is summarised in Fig. 8.28, as a guide when balancing the distribution patterns of new installations.

Similarly, velocities through return air grilles are related to the comfort of room occupants. The values shown in Fig. 8.29 are typical for wall or door-mounted return grilles.

Application	Max m/s	Max ft/min
Residential	2·54	500
Hotel bedrooms	3·80	750
Private offices	6·35	1 250
General offices	7·60	1 500
Shops	7·60	1 500
Factories	10·16	2 000

Fig. 8.28 Wall register outlet velocities

Grille location	max m/s	max ft/min
Close to occupants	1·52	300
High in side walls	2·54–3·05	500– 600
Ceiling returns	5·08–6·10	1 000–1 200

Fig. 8.29 Return grille velocities

The obvious item of test equipment with which to read air velocities is an air velocity meter as shown in Fig. 8.30. Note that instruments are also available which give direct readings of air *volumes*.

Balancing air distribution

The following sequence is usually satisfactory.

a) Check and accurately set the fresh air inlet and total (mixed) return air volumes. To do this use an air velocity meter over each 250 mm × 250 mm (10 in × 10 in) area of the return air inlet of the airhandling unit, then calculate the average the velocities to give the total flow through the entire inlet area.
b) Check the air velocity out of the evaporator fan discharge, working as above. Adjust the fan speed as necessary to secure design supply air volume.
c) Balance air volumes in main and branch ducts, working downstream, by adjusting volume or splitter dampers at the entry to each branch.
d) With all supply grilles wide open, and directional settings as shown in design drawings, work downstream from the start of each branch duct and set each supply grille or register to deliver the required volume. Use a velocity meter or a volume meter, whichever is available.
e) Set the return grille volume dampers,

Fig. 8.30 Air velocity meter

TROPICAL REFRIGERATION AND AIR-CONDITIONING

and check the air velocities in the return ducts.

f) Check that the entire conditioned space is free from draughts, stagnant areas and excessive noise levels; and that temperatures are equal throughout.

Note that small adjustments can be made to the settings of several dampers without upsetting the balance of a system as a whole. Similarly, evaporator fans can be reset within reasonable limits without upsetting the relative balance between ducts and diffusers. As a final precaution, check that the fusible links which hold open the heavy gauge steel blades of fire dampers are rated to rupture at the design temperature indicated on installation drawings. This is usually approximately 71 °C (160 °F).

Anti-vibration mountings and foundations

Much emphasis having been placed on the need to avoid transmitting noise or vibration, we must look briefly at various equipment and procedural options available.

Figure 8.32 illustrates a typical 'metallic hose' refrigerant line coupling. Designs for silver soldered joints are recommended, but flanged models are available. Similar types with synthetic rubber hosing are available for water lines.

Main plant must be isolated from floors (or ceilings) in most instances. Items such as airhandling units, which do not normally produce too much vibration, can be mounted on – or suspended from – either spring or rubber mounts or hangers.

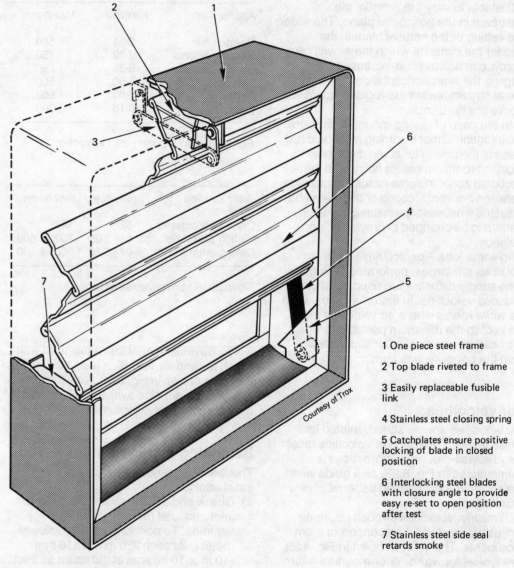

1 One piece steel frame
2 Top blade riveted to frame
3 Easily replaceable fusible link
4 Stainless steel closing spring
5 Catchplates ensure positive locking of blade in closed position
6 Interlocking steel blades with closure angle to provide easy re-set to open position after test
7 Stainless steel side seal retards smoke

Fig. 8.31 Fire damper layout

PIPELINE AND DUCTWORK INSTALLATION

Equipment generating more movement – pumps or compressors for example – are often mounted on steel bases. In the case of condensing units and packaged chillers, better quality plant incorporates isolators, and the bases need only be bolted down to prevent 'walking'. Alternatively, we can ourselves use structural rails or complete, site-fabricated steel bases as shown in Fig. 8.34(b). In either case, spring or rubber isolators must suit the load

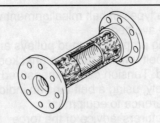

Fig. 8.32 Metallic hose refrigerant line connection

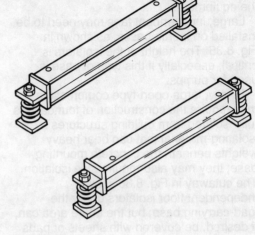

a) Rail type

b) Fabricated steel

Fig. 8.34 Bases

a) Spring

b) Rubber

Fig. 8.33 Isolators

distribution and operating characteristics of the equipment.

Large, independent fans may need to be installed on inertia bases, as shown in Fig. 8.35. The height of the isolators is critical, especially if this type of base is used for pumps.

Finally, large open-type compressors may require the construction of foundations fully isolated from building structures by isolating-material that can bear heavy weights beneath the concrete mounting base; they may also require side isolation. The cutaway in Fig. 8.36 shows independent floor isolators under the load-carrying base; but the entire area can, if desired, be covered with sheets or pads of isolating material. The important points are that this is thick enough and strong enough to carry the full weight of equipment and base and that the base is strong enough to support the machinery throughout an extended life. We rarely think in terms of reinforced concrete bases or floating floors less than 100 mm (4 in) thick; and 50 mm (2 in) of vibration-absorbing material is typical.

Drive kit installation

Vee belt drives

The most simple and effective aid to correctly aligning the pulleys of a motor and a compressor/pump/fan is a good metal straight edge, applied as shown in Fig. 8.37. The method ensures that the drive shafts as well as the pulleys are properly aligned, i.e. parallel. Figure 8.38 illustrates, in exaggerated form, three possible types of belt misalignment which must be avoided.

Having properly aligned pulleys and shafts, belt tension must be checked and adjusted. Tension can be measured accurately, using a belt tension indicator, with reference to equipment manufacturers' advice of the force necessary to deflect belts through a distance of 0·016 mm per 1·0 m span between the two shaft centres. The size of the required force will of course vary with the type and size of vee belts and pulleys. It will probably be somewhere between 41 and 82 kPa (6 and 12 lbf/in^2). If you know the required force but do not have a tension indicator of the type described, a spring balance can be used instead! Some motor mounting bases enable tension to be finely adjusted by turning a single bolt.

To balance fan or other equipment speeds, adjustable motor pulleys are often used. The pulley is made in two parts, which screw together and are locked by a grub screw. With the grub screw slackened off, the pulley section which is not locked to the motor shaft can be turned to increase or reduce the pitch diameter. This has the same effect as fitting either a larger or a smaller diameter non-adjustable pulley, and is of great value to installation and commissioning teams.

Basic motor pulley diameter requirements can be calculated using the formula:

$$\text{Motor pulley diameter} = \frac{\text{Fan r.p.m.} \times \text{Fan pulley diameter}}{\text{Fan motor r.p.m.}}$$

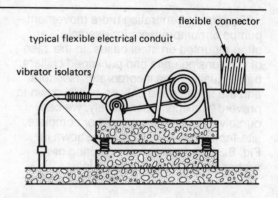

Fig. 8.35 Inertia base section

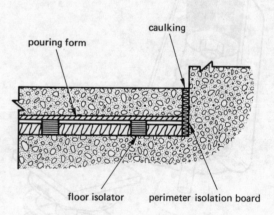

Fig. 8.36 Foundation section

PIPELINE AND DUCTWORK INSTALLATION

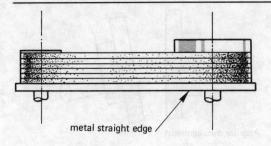

Fig. 8.37 Pulley alignment

Courtesy of Hall-Thermotank Products

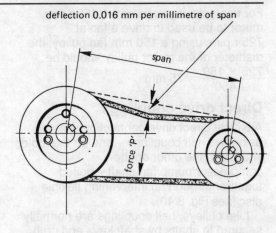

a) Measurement

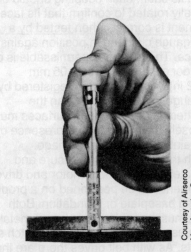

b) Indicator

Courtesy of Airserco Manufacturing Company

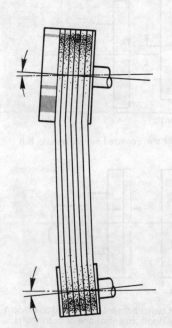 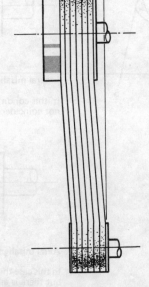 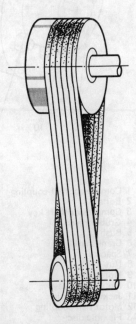

shafts are not parallel to one another | shafts are parallel and in alignment but the pulleys are not in alignment | shafts are not in correct alignment although they appear parallel when seen from above

Fig. 8.38 Vee belt misalignment

Fig. 8.39 Belt tension

161

TROPICAL REFRIGERATION AND AIR-CONDITIONING

For example, if you have a 1 450 r.p.m. motor to be used to drive a fan at 725 r.p.m. using a 150 mm fan pulley, the diameter of the motor pulley should be:

$$\frac{725 \times 150}{1\,450} = 75 \text{ mm}$$

Direct drive couplings

A typical direct drive connection is made using two half couplings – one on the motor shaft, and the other on the compressor/pump/fan shaft – bolted together through an intervening flexible disc (see Fig. 8.40).

Like pulleys, half couplings are normally secured to shafts by shaft keys and grub screws. In some cases they incorporate a central bush. When firmly secured at the end of its shaft, each coupling should be manually rotated to confirm that its face alignment is correct, when tested by a clock gauge clamped in position against the face. The tolerance permissable is of the order of plus or minus 0·05 mm (0·002 in). If the deflection registered by the gauge varies by more than the permitted value, all mating surfaces must be closely examined for the presence of dirt, paint or mechanical damage.

With the half couplings secure and accurately aligned, both motor and driven equipment can be positioned on a properly levelled baseplate or foundation. Both shafts must be in identical planes, metal shims usually being required to match shaft heights. As with belt drives there are three possible types of misalignment, illustrated in Fig. 8.41.

Recommended procedures are: first properly position and lightly secure the

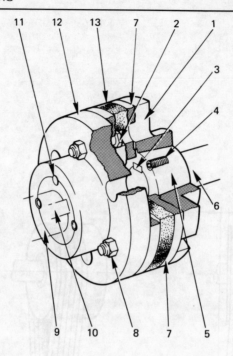

1 Compressor half-coupling
2 Bolts
3 Compressor shaft key
4 Grub-screws
5 Bush
6 Crankshaft
7 Spacers
8 Nut
9 Motor shaft
10 Motor shaft key
11 Jacking-off hole
12 Motor half-coupling
13 Flexible disc

Fig. 8.40 Direct drive coupling

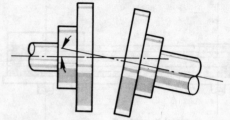

Angular misalignment

In this condition the shaft centre lines meet at a point midway between the shaft ends.

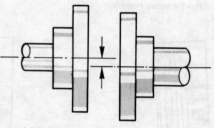

Lateral misalignment

In this condition the shaft centre lines are parallel but not coincident.

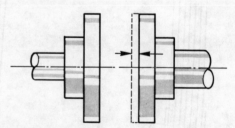

Axial misalignment

In this case the shaft centre lines are parallel and coincident but there is an axial displacement of one or both shafts. Remember that the gap between the coupling halves must be set with the compressor crankshaft on its front thrust face and the motor shaft on its magnetic centre.

Fig. 8.41 Half shaft misalignments

PIPELINE AND DUCTWORK INSTALLATION

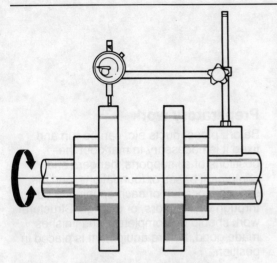

Fig. 8.42 Half coupling alignment checks

electric motor; place the equipment to be driven on the base, and balance shaft heights; then slide it forward until the distance between the two half couplings equals the thickness of the flexible disc. It also can then be lightly bolted down, and correct alignment of the perimeters of the half shafts secured and confirmed using a firmly clamped clock gauge. The usual tolerance, reading at several positions round each half coupling, is plus or minus 0·125 mm (0·005 in). When this is met, the flexible disc can be inserted, and the two half couplings bolted securely together. Coupling alignment tolerances must then be rechecked before the motor and compressor/pump/fan are finally bolted down. *Small* longitudinal corrections can be made by slackening off and moving one half coupling, but under no circumstances should flexible discs be forced between faces which are too close together. Before the motor can be started it is also necessary to confirm that compressor crankshafts are in any position (normally forward-thrust) specified by manufacturers, and that the entire drive can be rotated easily by manual pressure.

Revision questions

1. a) Explain the term 'non-condensible gases'.
   b) List two of the various causes of moisture in a refrigeration system and three of the effects which the entry of moisture is likely to have on a refrigeration system.
   c) State two precautionary measures that could be recommended to prevent the entry of moisture in the refrigeration system.
   (WAEC)

2. a) Describe how to open *and* how to close the service valves fitted to an open-type compressor.
   b) Detail the procedure to be followed when fitting a compound-type pressure gauge to such a valve and then using it.

3. Make a labelled line sketch showing the refrigerant lines connecting the main components of a charging station and a domestic air-conditioner. Show all the necessary shut-off valves.

9. Installation, commissioning and maintenance planning

In earlier chapters we have described specific installation and commissioning procedures; but only in the case of domestic appliances – especially room air-conditioners – have all needs been presented in a fashion which enables us to plan all operations in correct sequence. In the case of central plant, the fine detail of such procedures can only be established from technical data issued by individual manufacturers; but we should, and can, plan a systematic and logical procedural framework to hold the exact details applicable to most types of application.

Installation procedures

Site familiarisation
Before the equipment layout can be finalised, it is necessary to become familiar with the site, and the proposed locations of items of main plant. It may be necessary to make special arrangements to have heavy equipment hoisted and positioned on prebuilt bases. This may need to be done overnight or during a weekend, to avoid disrupting traffic. It is essential to ensure that plant will *not* obstruct emergency exists or passage ways, or block fire doors. It must be so positioned and isolated as to prevent annoyance to the occupants of this or adjacent buildings from noise and vibration. As well as the building structure, we must inspect and establish the suitability of power, water and drainage services and arrange any changes with the appropriate authorities, unless officially informed that this will be done by the owner, his architects, or other professional advisers. These points constitute aspects of contract management; but installation engineers must be familiar with needs and able to assess the information with which they are provided before site work commences.

Local regulations
It is of course essential that all materials and procedures comply fully with the requirements of relevant codes and regulations. The finished installation will probably need to be examined and approved by representatives of the Fire Brigade, the Electricity Supply Authority and the Water Department; not to mention factory inspectors and surveyors acting for an insurance company. Their individual and collective requirements must be established before designs of major installations can be completed; and those installing equipment must also be sure that their work will be in all respects acceptable. The final objective is to avoid accidents and hazards to personal safety of the building's occupants; and there are no acceptable shortcuts.

Preparatory work
Before pipes, ducts etc. can be run and fixed it is necessary to mark out the positions of all supports, hangers and equipment bases or foundations, and the size and location of each opening to be cut through walls, floors, or ceilings. Structural work should be completed, and finishes made good, before equipment is placed in position.

Transit damage
All equipment must be examined *immediately* it is delivered to site; arrangements should be made for any transit damage to be notified to carriers or suppliers in accordance with their terms of contract and, if necessary, used as the basis of insurance claims. There is always a strict time limit for the submission of claims – tomorrow can easily be too late!

Do not forget that plant can be damaged *after* arrival on site. Where possible, use sections of the original packing to protect against superficial damage.

Installation checklist
a) Locate all main equipment on prepared or anti-vibration mountings. Make sure that it is properly secured, and level in both planes. Use metal shims where height adjustments are required.

INSTALLATION, COMMISSIONING AND MAINTENANCE PLANNING

b) Slacken off or remove, as applicable, hold-down bolts and/or transit packing pieces.
c) Leak test (and where necessary repair and recharge) all refrigeration circuits.
d) Run and fix external refrigeration lines, as detailed in Chapter 8.
e) Connect water supply and/or drainage lines, and check strainers. Charge with water, adding water treatment chemicals. Purge all air from circuits. Start and check pumps. Balance water distribution and flow rates.
f) Connect supply and return air ducts, and fit or adjust grilles and registers. Check that air filters are in place. Start fans at a low speed and blow out dust or building debris from ducts. Finally, set, align and tighten fan drive components. Set correct volumes and balance air distribution as detailed in Chapter 8.
g) Make and check all electrical connections; check fuses, isolators, contacts etc.
h) Clean and tidy up all areas before proceeding further.

Pre start-up checks

a) Confirm that all fans are tight on shafts, rotate easily and turn in the correct direction.
b) Check the alignment of all drive kits, and that pulleys etc. are tight on shafts. Check belt tension.
c) Make sure that all bearings etc. are clean and properly lubricated.
d) Using control circuit current only, verify the operation (and sequences of operation) of all motor starters, controls and safety devices with reference to detailed wiring diagrams.

Start-up procedures

Follow detailed instructions provided by equipment manufacturers, or system designers. In general:
a) start all motors manually, in the correct sequence;
b) before starting compressors, run and test
  i) condensers (including towers, pumps, fans etc. as applicable) and the operating and safety controls in the condenser circuit,
  ii) evaporator fans (plus any accessories fitted) and air-side operating and safety controls;
c) make *certain* that compressor suction and discharge valves, and liquid-line valves as fitted are all *open* before compressors are started. Where possible, check and balance system operating pressures. Check the operation of all operating and safety controls.

Post start-up checks

a) Starting and running amperages of all motors.
b) Refrigerant suction and discharge pressures/temperatures, and oil pressures when instrumentation permits.
c) Air temperatures onto and leaving evaporator (or, where applicable, water or other fluid temperatures).
d) Chilled water temperatures into and out of evaporators and cooling coils.
e) Condenser air or water temperatures into and out of the condenser.
f) DB and WB temperatures of the ambient air.
g) DB and WB temperatures in the conditioned space.
h) Any abnormal noise or vibration.
i) Conditioned space at design conditions and free from draughts or stagnant air pockets, stratification etc.
j) All ancillary equipment and complete control circuits functioning correctly.
k) Log sheets and hand-over documentation complete.
l) Arrangements for any hand-over tests initiated.

Log sheets

It is desirable that all main operating characteristics – pressures and temperatures – be recorded at set times, at least twice per 24 hours. The information should be kept on well protected loose-leaf log sheets, or in bound log books. These should contain columns for time switched on, time switched off, and progressive totals of hours run by each compressor or pump with standby/changeover facilities. A 'Remarks' column 50–77 mm wide is also helpful, to record unusual events such as power failures, dust storms or other weather phenomena, and action taken outside normal maintenance procedures.

When entering log sheets, record what the instruments say, not the figures on the line above! The object of the exercise is to enable any deterioration in efficiency to be seen at an early stage, to help diagnose possible causes, and to trigger remedial action *before* major faults develop. The entry of full, clear and accurate information is one of the most important duties of plant operators.

It is not practical to illustrate a specimen log sheet which would cover all needs, in view of the diversity of the range of equipment in which we are interested. The sheet size and layout must be tailored to suit the equipment and the individual installation, just as the equipment and installation are designed to meet the needs of individual end users. In general, however, we need columns for each of the following:

a) date, and time;
b) ambient DB and WB temperatures;
c) space DB (plus WB where applicable, or RH);
d) compressor suction and discharge pressures, oil pressure, running amperage;
e) temperatures of air or other fluid entering and leaving each main cooling coil;
f) temperatures of chilled water or other secondary refrigerants entering and leaving each chiller;
g) temperatures of air or water entering and leaving each condenser;
h) where applicable, temperatures of air and of water entering and leaving each water cooling tower or evaporative condenser, including the WB reading;
i) running amperage of each major fan, pump or other motor either vital to the installation or having a capacity exceeding 15 kW;
j) the remarks to record everything out of the ordinary which happens during the course of the day.

Interpretation of log sheets

Unusual weather conditions will often have an early and direct effect on the figures logged. A sand storm might, for example, cause partial blockage of of the space between air cooled condenser fins; if so, condensing pressure and compressor motor running amperages will both increase. This is a clear indication that the condenser surfaces must be cleaned as soon as conditions permit. If the sand storm is noted in the remarks column, it will also act as a reminder that the condition of all air filters must be checked. If the evaporator filter becomes very dusty, airflow over the coil will be restricted. The refrigerant suction temperature will fall, and room temperatures rise.

These symptoms could of course have other causes; but if the sand storm has been noted in the log, it will not be necessary to complete a detailed check through the entire installation in search of obscure defects. So long as readings return to normal as soon as the plant has been cleaned up, the entry will have provided an acceptable explanation of the variations in performance details.

Planned maintenance

The exact details of maintenance operations, and timetables, will vary with individual installations and local operating conditions. Manufacturers issue their own detailed maintenance instructions which should be collated to form fully detailed schedules. These might include the following.

Refrigeration

Daily
– Check that all equipment is free from noise and vibration.
– Check installation operating temperatures at 0800, 1230 and 1700 o'clock.
– Check suction, discharge and oil pressures.

Weekly
– Check the liquid-line sight glass.
– Check running amperages of all main motors.
– Check that air-cooled condensers are free from obstructions.

Monthly
– Check all belts for condition, tension and alignment.
– Check compressor oil levels (if a crankcase sight glass is fitted).

Quarterly
– Check water-cooled condensers for levels of chemical treatment.
– Check finned heat exchanger surfaces for freedom from blockages.
– Lubricate motor/fan bearings as necessary.
– Leak test the refrigeration system, using an electronic leak detector.

Yearly
– Examine all motor/fan/shaft bearings.
– Clean all water strainers.
– Empty, clean and repaint as necessary all cooling tower/condenser surfaces exposed to water, renew water treatment chemicals.
– Check *all* operating and safety controls for correct setting and operation.

Air-conditioning

Requirements will be basically as listed above, but the following additions are necessary:

Weekly
- Check/clean/replace air filters as necessary.
- Check the condition of air intake grilles and filters.
- Check that condensate drains are free from blockage.

Quarterly
- Wash out condensate trays etc. with a biocide.
- Inspect humidifiers, if fitted, for freedom from scale.
- Inspect and if necessary clean sensors of thermostats and humidistats.

6-monthly
- Re-activate sensing elements of hair-type humidistats in accordance with manufacturers' instructions.

Special equipment

When installations use either centrifugal or absorption-type water chillers, the purge system should be operated weekly to remove air, moisture, etc. from systems. Quarterly programmes should include the lubrication of external capacity control motors and linkages. The linkages and stems of water control valves should be cleaned and lubricated not less than twice a year.

Although it is not often given prominence in manufacturers' literature, *cleanliness* is one of the most important requirements, especially in the case of controls, and electric contacts. These must be inspected regularly to ensure not only that they remain free from dust or other contaminants, but also that sensors are not screened from the element whose condition they control, or affected by external sources of heat or cold.

Things to avoid

It is a tradition in many branches of engineering that equipment is stripped, examined and reconditioned at fixed intervals, as part of planned maintenance operations.

It is essential that this practice is *not* followed when maintaining refrigeration systems, which should only be opened and exposed to atmospheric conditions when it is necessary to repair or replace a component. Ideally, a refrigeration system should have an operating life of – according to type and application – between 10 and 25 years. During that time, a number of routine maintenance operations will be necessary to prevent mechanical or electrical overloads; but the chances are that it will not be necessary to repair or replace an internal component during the working life of the equipment. If it is, every care must be taken to avoid introducing contaminants which might seriously damage mechanical and/or electrical components. This observation may seem superfluous, but the authors have seen boiler maintenance procedures, for example, applied to refrigerant condensers! Remember that you are specialising on equipment with its own individual characteristics and needs; and that all maintenance operations must be carried out in accordance with the procedures detailed in Chapters 11 and 12.

Revision questions

1. An R12 commercial condensing unit is used to supply liquid to two evaporators working at different conditions. If the evaporating temperatures are $-5\,°C$ and $-15\,°C$:
   a) draw the refrigerant circuit diagram,
   b) insert in the diagram necessary controls for the efficient operation of the unit,
   c) suggest suitable settings for the cut-in and cut-out pressures of a low pressure switch used to control the compressor.
   (WAEC)

2. Show the layout of a page of a log book to be used when maintaining the refrigeration equipment in a block ice making plant, incorporating an R717 compressor, a shell-and-tube condenser, a water cooling tower and a flooded-type submerged evaporator.

10. Trouble-shooting

This chapter summarises in tabular form the symptoms of defects most likely to be encountered when servicing domestic and commercial refrigeration and air-conditioning equipment; lists possible causes and appropriate remedial action; and refers to other chapters or pages having a direct bearing on the problem. It cannot cover all possible service problems and should, whenever possible be read in conjunction with service manuals for specific items of equipment. To avoid repetition, subject matter has been split into sections, which are not all self-contained – for example, a number of symptoms listed under 'Airhandling Units' could occur on forced draught evaporators in walk-in coldrooms and also on air-conditioners. The sections are individually numbered, indexed, and cross-referenced. Sections 10.1 and 10.5–10.15 have drawn on Airedale's Installation and Maintenance Manual. **The safety precautions detailed in specific chapters on electric and refrigeration system service procedures must at all times be observed. Many of the operations listed require specialist knowledge and experience, and should be carried out only by fully trained service engineers and/or electricians.**

10.1 Whole System (Electrical)

Symptom	Possible cause	Remedial action	Refer to
A) No motors start – no power at mains	a) Mains fuses blown	a) Contact Electricity Board; locate and remedy cause of overload if inside building	
	b) Wiring defect external to equipment	b) Check switchgear and cables/connections back to mains distribution board; adjust or replace defective items	
B) No motors start – power reaching mains disconnect switch	a) Unit fuses blown	a) Replace fuses; locate and remedy cause of overload	Ch. 12
	b) Unit wiring fault	b) Check all controls and switchgear, cable and connections for condition and correct wiring	Ch. 12
	c) Controls not properly set	c) Check and reset operating controls	Ch. 12
	d) Evaporator fan motor or starter defect	d) Check both items and replace if necessary	10.19

TROUBLE-SHOOTING

10.2 Self-contained Mechanical Refrigerators

Symptom	Possible cause	Remedial action	Refer to
A) Power at point but unit will not start	a) Thermostat improperly set or defective	a) Reset or test and replace thermostat	Ch. 5
	b) Compressor overload open circuit	b) Test, repair or replace Klixon	Ch. 5
	c) Wiring defect	c) Check wiring for connections and continuity; adjust as necessary	Ch. 12
B) Compressor hums and trips on overload	a) Low voltage	a) Check voltages at power point and at terminal box	
	b) Defective capacitor	b) Test and replace if necessary	Ch. 12
	c) Defective relay (not closing)	c) Check relay operation and replace if necessary	Ch. 5 / Ch. 12
	d) Defective compressor	d) Check for mechanical operation – replace if piston etc. is jammed	Ch. 8 / Ch. 11
	e) Liquid refrigerant in compressor	e) Fit crankcase heater and/or accumulator	Ch. 5
C) Compressor short-cycles	a) Defective compressor overload	a) Test, repair or replace Klixon	Ch. 5
	b) Thermostat defective	b) Test, replace if necessary	Ch. 5
	c) Condenser airflow restricted.	c) Check cleanliness of condenser – remove any obstructions to correct airflows	Ch. 5
	d) Air in system	d) Test, purge, repair leak and recharge system	Ch. 8 / Ch. 11
	e) Liquid-line damage	e) Check liquid line for damage and replace if necessary	Ch. 8
	f) Liquid-line filter blocked (filter frosts over)	f) Replace strainer/drier and capillary tube	Ch. 6 / Ch. 11
	g) Capillary tube blocked or damaged (frost back along capillary possible)	g) Replace capillary tube and strainer/drier	Ch. 5 / Ch. 8 / Ch. 12

TROPICAL REFRIGERATION AND AIR-CONDITIONING

10.2 Self-contained Mechanical Refrigerators

Symptom	Possible cause	Remedial action	Refer to
D) Cabinet temperature high, compressor runs continuously	a) Shortage of refrigerant	a) Leak test, repair leaks if present, recharge system	Ch. 8
	b) Condenser efficiency low	b) Check condenser for cleanliness and remove any obstruction to correct airflow	Ch. 5
	c) Compressor valve defect	c) Check pumping efficiency, replace if necessary	Ch. 11
	d) Door distorted or door gasket defective (evaporator heavily coated in moist ice, condensate around door)	d) Check door and gasket seals – replace if necessary	Ch. 11
	e) Cabinet overloaded with hot foodstuffs or liquids	e) Arrange for proper loading procedures to be followed	
	f) Air in system (condenser and compressor hot)	f) Purge air, leak test, repair leaks and recharge	Ch. 8 Ch. 11
	g) Blockage in capillary or liquid line	g) Check all components and if necessary replace them	
E) Noisy operation	a) Pipes or components rattling	a) Check, tighten any loose components; isolate or relocate pipes to prevent physical contact; check compressor external mountings	Ch. 8
	b) Loose/defective external compressor mountings	b) Check, and if necessary replace	Ch. 8
	c) Cabinet not level	c) Check levels, adjust as necessary	
	d) Defective compressor (valves, internal springs etc.)	d) Replace compressor if noise is internal	Ch. 11

TROUBLE-SHOOTING

10.3 Self-contained Absorption-Type Refrigerators

Symptom	Possible cause	Remedial action	Refer to
A) No or poor cooling efficiency	a) Gas jet burner or heating element defective	a) Strip, clean and check correct operation; replace wick etc. if required	
	b) Thermostat defective	b) Check operation, replace if necessary	Ch. 5
	c) Door gasket faulty or door not sealing properly (condensation around door, evaporator coated with watery ice)	c) Check door seal and gasket fit – replace as necessary	
	d) Loss of refrigerant charge (faint smell of ammonia)	d) Fit new cooling system	
	e) Poor condensing efficiency	e) Check condenser for cleanliness and remove any obstructions to free airflow	Ch. 5

10.4 Walk-in Refrigerators

Symptom	Possible cause	Remedial action	Refer to
A) Temperature high – coil heavily frosted	a) Defective door gasket (condensation around door)	a) Check, replace if necessary; investigate loading methods	
	b) Inadequate defrost cycle	b) Examine time clocks, heating elements and safety controls or reverse cycle defrost valves as fitted; if all is in order, check that coil is free from frost at end of cycle (time) and speed of ice build-up (frequency) and reset controls as necessary	Ch. 6
	c) Excessive loading	c) Investigate pattern of usage and control settings if produce is held near its freezing point	Ch. 6
For other symptoms see 10.5–8			

10.5 Airhandling Units – General

Symptom	Possible cause	Remedial action	Refer to
A) Evaporator fan motor runs – no cooling or heating	a) Controls not properly set	a) Check and reset	Ch. 12
	b) Defective thermostat	b) Check and reset	Ch. 5
	c) Starting sequence interlocks not completed	c) Locate and correct wiring or control defect	Ch. 12
	d) System wiring defect	d) Locate and remedy fault	Ch. 12
	e) Safety control open circuit	e) Locate and remedy cause of control operation	Ch. 6
B) Evaporator fan motor runs – insufficient cooling or heating	a) Blocked evaporator airflow	a) Check grilles for obstructions, condition of cooling coil and air filter; clean and adjust as necessary	Ch. 5
	b) Defective evaporator fan or fan drive	b) Check fan tight on shaft and rotates freely; inspect belt tension and condition; adjust or replace as necessary	Ch. 8
C) Excessive noise	a) Pipes vibrating and making metal to metal contact	a) Adjust pipe positions	Ch. 8
	b) Evaporator fan touching blower housing	b) Reposition and tighten fan(s) on shaft	Ch. 8
	c) Fan shaft out of alignment	c) Test, replace if necessary	Ch. 8
	d) Fan drive defect	d) Check belts for wear, tension and alignment; check bearings and pulleys	Ch. 8
	e) Fan out of balance	e) Rebalance or replace	
	f) Anti-vibration isolators worn or defective	f) Replace as necessary	Ch. 8
	g) Faulty installation – unit not properly level	g) Check levels, adjust or correct as necessary	

10.6 Airhandling Units – Direct Expansion

Symptom	Possible cause	Remedial action	Refer to
A) Evaporator fan runs compressor will not start	a) Compressor electrical interlocks not completed	a) Locate and remove cause	Ch. 12
	b) Compressor safety control open circuit	b) Locate and remove cause	Ch. 5 Ch. 12
	c) Defective compressor	c) Check, and replace if necessary	Ch. 12
B) Evaporator fan runs – insufficient cooling – compressor runs continuously	a) Controls not properly set	a) Check and reset	
	b) Shortage of refrigerant	b) Check via sight glass – top-up charge, leak test	Ch. 8
	c) Compressor not pumping	c) Test, replace if necessary	Ch. 11
	d) TEV jammed open	d) Strip, clean or replace	10.7 Ch. 6
	e) Excessive heat load	e) Check for structural or equipment causes	
	f) Flash gas in liquid line – low ambient temperature	f) Fit head pressure controller	Ch. 6
C) Evaporator fan runs – insufficient cooling – compressor short-cycles	a) Low voltage	a) Check voltage at mains isolator and compressor terminals with unit running	
	b) Controls not properly set	b) Check and adjust	
	c) Air in system	c) Purge, repair leak, top up refrigerant charge	Ch. 8 Ch. 11
	d) Inadequate condensing	d) Check condenser fan operation and for freedom from blockages	
	e) Defective compressor contact or windings	e) Check connections and operation – replace parts as necessary	Ch. 11
	f) Compressor safety control open circuit	f) Check all control settings and operation, or locate and remedy cause if external	Ch. 6
	g) Inadequate evaporator airflow	g) Check for blockages of air filters, cooling coil, air passages; check condition of fan and drive kit	
	h) Liquid-line blockage	h) Check condition of filter/drier, solenoid valve and TEV; check that TEV phial is properly secured and has not lost its charge	Ch. 6 10.4 10.8
	i) 'Over condensing'	i) Fit head pressure controller	Ch. 6
	j) Shortage of refrigerant	j) Check at sight glass, leak test, repair leaks if found, top-up refrigerant	Ch. 8

10.6 Airhandling Units – Direct Expansion

Symptom	Possible cause	Remedial action	Refer to
D) Low suction pressure	a) Expansion valve defect – seat or superheat setting	a) Systematically check TEV operation, clean or replace or adjust as necessary	10.7
	b) Evaporator airflow incorrect	b) Check fan tightness, drive kit condition, absence of restrictions to airflow	Ch. 8
	c) Shortage of refrigerant	c) Check at sight glass, leak test, repair leaks, top up with refrigerant	10.7
	d) 'Over condensing' – low ambient temperature	d) Fit head pressure controller	Ch. 6
	e) Incorrect control setting	e) Check control settings and operation	
	f) Partial obstruction of liquid line	f) Check condition of filter/drier and solenoid valve	10.8
	g) Flash gas in liquid line – no leaks in system	g) Check height of liquid-line risers and degree of subcooling in condenser	Ch. 6

10.7 Thermostatic Expansion Valves

Symptom	Possible cause	Remedial action	Refer to
A) Low suction pressure – high superheat	a) Low ambient temperature b) Loss of refrigerant c) Equaliser line blocked d) Orifice blocked by wax, oil or dust e) Superheat setting too high f) Partial loss of sensing phial charge	a) Fit head pressure control b) Trace and repair leak, recharge c) Repair or replace line d) Strip and clean valve; fit new filter/driers e) Adjust setting f) Replace valve power element	Ch. 6
B) Low suction pressure – low superheat	a) Incorrect refrigerant distribution	a) Check condition of distributor	
C) High suction pressure – low superheat	a) Incorrect superheat setting b) Valve seat held open by dirt, oil etc. c) Equaliser line blocked d) Bulb not properly fixed	a) Adjust setting b) Clean valve, fit new filter/drier c) Clean or replace line d) Check that bulb is securely clamped to suction line and is not affected by other heat sources	
D) Suction pressure fluctuates ('Hunting')	a) Incorrect superheat setting b) Uneven load at evaporator c) Equaliser line restricted d) Oil in suction line affecting temperature sensed by valve bulb	a) Adjust setting b) Check condition of distributor c) Clean or replace line d) Do not locate valve sensing element in a section of line likely to form an oil trap	

10.8 Solenoid Valves

Symptom	Possible cause	Remedial action	Refer to
A) Valve will not open	a) Plunger stuck by solids or oil b) Valve body warped c) Solenoid coil burnout d) Faulty electric connections e) Excessive refrigerant pressure	a) Strip and clean valve b) Replace body c) Remedy cause and replace spring d) Check all contacts and electric connections e) Locate and remove cause	Ch. 6
B) Valve will not close	a) Plunger stuck by solids or oil b) Spring broken or stuck c) Electrical fault	a) Strip and clean valve b) Strip valve, replace spring c) Check all contacts and electric connections	
C) Valve closes but flow continues	a) Contaminants under valve seat b) Seat or pin damaged	a) Strip and clean b) Replace damaged item(s)	
D) Valve noisy	a) Incorrect assembly b) Refrigerant noise c) Flash gas in liquid line d) Electric hum	b) Check and tighten components b) Install discharge muffler c) Check refrigerant charge and liquid-line strainer/drier; check whether head pressure control is necessary d) Check coil sleeve for fit and overall valve cleanliness	
E) Coil burned out	a) High or low voltage b) Wiring defect c) Moisture entered coil d) Plunger jammed	a) Check and remedy b) Check all contacts and electric connections c) Protect from drips and electrically seal all connections d) Strip and clean valve when fitting new coil	

TROUBLE-SHOOTING

10.9 Chilled-Water Airhandling Units

Symptom	Possible cause	Remedial action	Refer to
A) Insufficient cooling	a) Insufficient chilled water supplied to cooling coil	a) Check setting and operation of controls, and chilled water valve(s)	Ch. 8 10.18
	b) Chilled-water flow switch open circuit – cooling system inoperative	b) Locate reason for operation of safety control, and restart system; if no other defect is apparent, check the flow switch itself	
	c) Restricted airflow over cooling coil	c) Systematically check through air filter condition, and the cooling coil cleanliness; check fan operation, belt tension and alignment, and shaft bearing condition	
	d) Excessive cooling load	d) Check conditioned space for open doors, windows etc; or for the use of additional heat generating equipment	
	e) Internal scaling of chilled water system	e) Test pressure drops through cooling coil and water chiller; if it is excessive, clean entire circuit chemically; recharge chilled water system and add a charge of a scale-inhibiting chemical	Ch. 8
B) Overcooling of space	a) Incorrect control setting	a) Check and reset as necessary with particular reference to thermostat, water valve linkage and valve operation	
	b) Excessive chilled water flow rate	b) Check controls as at (a) above; test water volumes through coil and if necessary adjust and rebalance system water distribution	Ch. 8

TROPICAL REFRIGERATION AND AIR-CONDITIONING

10.10 Airhandling Units – Electric Heating Coils

Symptom	Possible cause	Remedial action	Refer to
A) Insufficient or no heating effect	a) Controls not properly set	a) Check and reset	
	b) Control defect	b) Systematically check contacts and operation of operating controls; adjust or replace items as required	Ch. 12
	c) Restricted airflow over heating coils	c) Check air filters, cooling coil, supply and return air grilles for freedom from obstructions	Ch. 11
	d) Defective heating elements	d) Check all elements for electrical continuity and replace where necessary	Ch. 12
	e) Heater safety switch faulty or open circuit	e) Check safety switch for correct operation; if this is in order check out and remove cause of operation; if defective, replace safety switch	Ch. 11
	f) Unit wiring defect	f) Check conductors, connections etc. against wiring diagram	Ch. 11
	g) Evaporator fan or fan drive defect	g) Check fan tight on shaft and rotates freely; check tension, alignment and condition of fan motor belts, and tightness of pulleys; check bearings	Ch. 8
B) Overheating of space	a) Controls not properly set	a) Check setting of thermostat etc.	Ch. 11
	b) Controls defective	b) Systematically check contacts and operation of thermostat; adjust or replace as necessary	

TROUBLE-SHOOTING

10.11 Airhandling Units With Humidifiers

Symptom	Possible cause	Remedial action	Refer to
A) Humidity too low	a) Controls not properly set	a) Check and adjust	
	b) Defective control	b) Check control circuit operation and condition of electric contacts, replace as necessary	Ch. 7
	c) Excessive scale in humidifier	c) Replace or descale, in accordance with instructions for the specific type used	Ch. 7
	d) Humidifier undersized for operating conditions	d) Recalculate needs and if necessary arrange for a booster humidifier to be fitted	
	e) System wiring defect	e) Check all wiring, contacts and fuses	Ch. 7
	f) Cooling system defect	f) Check air temperature off cooling coils; if it is too low proceed as instructed, in section on direct expansion or chilled-water coils, as applicable	
B) Humidity too high	a) Controls not properly set	a) Check and adjust	
	b) Defective control	b) Check control operation, and state of electric contacts; replace as necesary	Ch. 12

179

TROPICAL REFRIGERATION AND AIR-CONDITIONING

10.12 Condensing Units (General)

Symptom	Possible cause	Remedial action	Refer to
A) Compressor motor will not start	a) No power at terminals	a) Check mains supply, fuses, control settings and terminals, and interlocks	Ch. 12
	b) Power arriving at motor terminals	b) Check pressure and thermal overloads – adjust them, or trace and remove cause of operation; test motor insulation and winding continuity, and freedom from mechanical seizure; if necessary, replace compressor	Ch. 12
B) Compressor short-cycles	a) Incorrect control or safety device setting	a) Check and reset as necessary	
	b) Restrictions in liquid line	b) Check solenoid valve, TEV, filter/drier and piping itself	10.7 10.8
	c) Shortage of refrigerant	c) Check at sight glass, top-up charge, leak test, repair leaks	Ch. 8
	d) Refrigerant overcharge	d) Check discharge pressure variations during off-cycle, purge any air from high side of condenser	Ch. 11
	e) Insufficient cooling duty	e) Investigate load at evaporator; if necessary fit hot gas bypass system	10.15
	f) Over condensing – flash gas in liquid line preventing proper start-up	f) Check temperature of air onto (or water into) condenser, and cooling requirements; if necessary fit head pressure controller	Ch. 6
C) Compressor runs non-stop	a) Excessive cooling duty	a) Check and if possible remove cause of excess load (open doors or windows etc. or use of extra heat-producing appliances)	
	b) Incorrect operating control settings	b) Check and reset as necessary	
	c) Defective compressor refrigerant suction/discharge valves	c) Test compressor pumping efficiency and maintenance of pressures during off-cycle; replace if necessary	Ch. 11
	d) Faulty refrigerant flow	d) Check for leaks, defective TEV operation or superheat setting, and phial fastening; check solenoid valve operation and filter/drier condition	10.7 10.8
D) Suction line frosts back to compressor	a) Insufficient cooling duty	a) Investigate load at evaporator; if necessary fit hot gas bypass system	

10.12 Condensing Units (General)

Symptom	Possible cause	Remedial action	Refer to
E) Compressor excessively noisy	a) Faulty external mountings	a) Check, replace as necessary	
	b) Unit jumped off internal spring mountings	b) Replace compressor	Ch. 8 Ch. 11
	c) Damaged valve reeds etc.	c) Replace compressor	
	d) Liquid slugging	d) Check and adjust TEV superheat setting, and cooling duty at evaporator; check evaporator free from restrictions	10.7
F) Head pressure too high	a) Overcharge of refrigerant	a) Purge off excess	
	b) Air in system	b) Verify through check of pressure during off-cycle; purge air	Ch. 11
	c) Inadequate condenser performance	c) Check flows of air or water and ensure that condenser cooling surfaces are clean	
G) Head pressure too low	a) Shortage of refrigerant, or restricted liquid line	a) Check at sight glass; and condition of strainer/drier, solenoid valve and TEV	10.7 10.8
	b) Insufficient cooling load	b) Check condition of evaporator and air filters in DX (direct expansion) units	
	c) Over condensing	c) Check entering temperature of cooling medium; if necessary fit hot gas bypass system	10.15
	d) Hot gas bypass valve wrongly set	d) Check and adjust	10.15
	e) Defective compressor discharge valve	e) Check pumping efficiency, if necessary replace	Ch. 11
H) Suction pressure too high	a) Excessive cooling load	a) Check evaporator duty	
	b) TEV setting or operation faulty	b) Check, adjust or replace TEV as necessary	10.7
	c) Defective compressor suction valve	c) Check pumping capacity and replace if necessary	Ch. 11
I) Suction pressure too low	a) Shortage of refrigerant, or liquid-line restriction	a) Check at sight glass; then if necessary conditions of filter/drier, solenoid valve and TEV; top up, adjust or replace as necessary	10.8 10.7
	b) Hot gas bypass valve not correctly set	b) Test and adjust as necessary	10.15
	c) Excessive pressure drop through evaporator	c) Check and adjust TEV superheat setting as necessary	10.7
	d) Restriction at evaporator of air or water to be cooled	d) Check air/water volumes over evaporator and performance/condition of fans, filters, pumps etc; clean or adjust as necessary	

TROPICAL REFRIGERATION AND AIR-CONDITIONING

10.13 Condensing Units (Air-cooled)

Symptom	Possible cause	Remedial action	Refer to
A) Low discharge pressure	a) Low ambient temperature	a) Check settings and operation of head pressure controller and dampers if fitted; adjust or replace as necessary	
B) High discharge pressure	a) High ambient temperature	a) As at (a) above	
	b) Condenser airflow restricted	b) Remove any obstructions to airstreams; check condenser fins not choked by dirt, clean as necessary	
	c) Condenser fan drive defect	c) Check condition and tightness of fan – belt tension, alignment and wear – freedom of shaft – motor operation	Ch. 8
	d) External influences	d) Check no short-cycling of air between adjacent units – no exposure to heat from other equipment	

10.14 Condensing Units (Water-cooled)

Symptom	Possible cause	Remedial action	Refer to
A) Low discharge pressure	a) Condenser water too cool, or flow rate too high	a) Check regulating valve settings and pump operation	10.18
	b) Low ambient temperature	b) If all system components are in good order cycle cooling tower fan(s) by fitting thermostat or pressure control	Ch. 6
	c) Pressure relief valve operated – shortage of refrigerant	c) Top up with refrigerant, check pressures, remove any cause of excessive condensing pressures	Ch. 6
B) High discharge pressure	a) Condenser water too hot, or flow rate too low	a) Check cooling tower performance and fan/drive/motor condition; check condition of water sump filter, and pump operation	10.17 10.18
	b) Condenser tubes scaled up	b) Check water flow rate and pressure drop. If these suggest scale is present, check visually and descale mechanically or chemically as convenient; add anti-scaling kit in cooling tower water system	10.17 Ch. 8

10.15 Hot Gas Bypass Valves

Symptom	Possible cause	Remedial action	Refer to
A) Valve will not open	a) Seat or plunger held by wax, dirt or oil	a) Strip and clean valve, including entry port strainer	Ch. 6
B) Valve will not close	a) Seat or plunger held by wax, dirt or oil	a) Strip and clean valve, including entry port strainer	
	b) Diaphragm fouled	b) Replace top element	
	c) Valve bypass passage blocked	c) Strip and clean valve, using R11 and pump if necessary; clean entry port strainer; fit new liquid-line strainer	
	d) External bypass line restricted	d) Replace with new line and check pressure drop is correct before leaving system	

10.16 Water Chillers

Symptom	Possible cause	Remedial action	Refer to
A) Compressor will not start	a) Safety control open circuit	a) Check all operating and safety controls and water flow switch if fitted	Ch. 7
	b) Control circuit not made	b) Check pumps etc. wired ahead of compressor motor are all operating properly	10.17
B) Chilled water temperature too high	a) Control setting incorrect	a) Check all operating and capacity controls	
	b) Condenser efficiency low	b) Check condenser free from scale, corrosion (water-cooled) or restrictions to airflow (air-cooled)	
	c) 'Over condensing'	c) Check head pressure controller and accessories	Ch. 6
	d) Chiller efficiency low	d) Check chilled-water system strainer, water treatment chemicals, and flow rates	Ch. 8
	e) Water cooling tower or evaporative condenser fault	e) See 10.17	
	f) Safety control open circuit	f) Check all operating and safety controls, and water flow switch if fitted	
C) Chilled water temperature too low	a) Incorrect control setting	a) Check all operating controls	
	b) Cooling load at airhandling units too low	b) Discuss with Design Department – capacity control provisions are probably inadequate	
	c) Restriction in water flow	c) Check operation of pump and all chilled-water control valves	
	d) Pump efficiency reduced	d) See 10.18	

TROUBLE-SHOOTING

10.17 Water Cooling Towers/Evaporative Condensers

Symptom	Possible cause	Remedial action	Refer to
A) Insufficient airflow	a) Fans loose, slack belts etc. b) Noise or vibration from fan drive c) Restricted airflow	a) Check condition of all drive components b) Check alignment and tightness of drive components c) Remove obstructions	Ch. 6 Ch. 8
B) Water level low	a) Make-up valve jammed shut b) Water supply interrupted c) Inlet water strainer choked	a) Clean and lubricate valve b) Locate and remove cause c) Strip and clean all strainers	
C) Water flow rate low	a) Condenser tubes scaled b) Pump operating defect c) Water circuit partially blocked	a) Inspect and clean all tubes; add full charge of water treatment chemicals; check pressure drop now correct b) See 10.18 c) Strip and clean *all* strainers; check pressure drop through system	
D) Water temperature drop through tower below correct figure	a) Condenser tube scaling b) Insufficient airflow c) Water spray nozzles choked d) Fill and/or cooling coil surfaces corroded or dirty	a) Inspect and clean tubes; add proper charge of water treatment chemicals b) Remove obstructions c) Remove and clean nozzles d) Strip and clean fill and descale cooling coil surfaces with chemical cleaner	

10.18 Water Pumps

Symptom	Possible cause	Remedial action	Refer to
A) Pump runs, no water flow	a) Pump not primed b) Air in circuit c) Clogged water strainer d) Blockage in pipeline e) Impeller rotation incorrect	a) Ensure circuit filled b) Purge all air c) Clean and replace d) Remove foreign matter e) Check and correct direction of rotation	
B) Water volume too low	a) Partial blockage in system b) Incorrect rotation direction c) Design fault – insufficient lift d) Mechanical wear	a) Check and clean b) Check and correct c) Refer to Design Department d) Examine and replace bearings and other components subject to mechanical wear	
C) Excessive noise	a) Shortage of lubricant b) Impeller out of balance c) Mechanical wear d) Drive misalignment	a) Oil all bearings and moving parts b) Examine fitting and clearances of mechanical components c) Examine and replace bearings and other components subject to mechanical wear d) Check and correct as necessary	
D) Vibration	a) Loose hold-down bolts b) Drive misalignment c) Bearings and/or shaft wear	a) Tighten as necessary b) Check and correct as necessary c) Check and replace as necessary	
E) Motor overheats	a) Mechanical – see 10.19	a) See Fig. 10.19	
F) Water leaks from seal	a) Drive misalignment b) Shortage of lubricant c) Mechanical wear	a) Check and correct as necessary b) Check and lubricate c) Replace seal	
G) Excessive shaft end play	a) Bearings worn b) Mechanical wear	a) Check, replace if necessary b) Check all wearing parts and replace as necessary	10.19 10.19
H) Loss of performance	a) Blocked water strainer or system b) Rust, scale, sand etc. in system c) Impeller out of balance d) Mechanical wear	a) Clean strainer/pipelines b) Clean strainer/pipelines c) Strip pump, check shaft, bearings and impeller for wear and correct assembly d) Check all wearing parts and replace as necessary	Ch. 8

10.19 Electric Motors

Symptom	Possible cause	Remedial action	Refer to
A) Loud humming noise	a) Uneven gap between stator and rotor, due to worn bearings b) Rotor out of balance	a) Replace bearings; check lubrication and drive kit alignment b) Replace motor	
B) Loud knocking noise	a) Shaft misaligned	a) Realign shaft. Examine and if necessary replace bearings etc.	
C) Steady tapping or clicking noise	a) Dirt or other foreign matter inside frame	a) Strip and clean motor	
D) Vibration	a) Rotor out of balance b) Shaft misaligned	a) Replace motor b) Realign shaft; examine and if necessary replace bearings etc.	
E) Bearings overheating	a) Drive belts or coupling out of alignment b) Shaft misaligned c) Shortage of lubricant d) Moisture/condensation entering frame	a) Check and adjust as necessary b) Realign shaft and if necessary replace bearings etc. c) Check and correct d) Remove cause or fit totally enclosed motor	
F) Motor overheating	a) Drive belts or coupling out of alignment b) Belt tension too tight c) Mechanical overload d) Dirt inside frame e) Rotor and stator in contact f) Low voltage g) Shortage of lubricant h) Excessive ambient or plantroom temperatures	a) Check and adjust b) Adjust as necessary c) Check condition of driven equipment; lubricate, re-align or adjust as necessary d) Strip and clean motor e) Check bearings, replace if necessary; check shaft alignment f) Check supply voltage at motor g) Check and correct h) Ensure adequate plantroom ventilation	

11. Refrigeration system service

For simplicity, service operations have been divided into two groups, which are covered in separate chapters. Chapter 11 deals with work on refrigeration systems, whilst electrical tests and procedures are covered in Chapter 12.

A number of relevant procedures have been described in earlier chapters, and will not be repeated. For example, Chapter 8 details procedures to be followed when cutting and jointing pipelines; brazing or silver soldering; evacuating and leak testing; and charging with either gaseous or liquid refrigerants. Noise and vibration precautions, and drive kit adjustment methods are also described in Chapter 8; and many service pointers are included in earlier chapters describing specific system components.

Safety precautions

Precautions to be taken when handling refrigerants or other chemicals, and flushing out or brazing pipelines which have contained refrigerants, have also been detailed earlier. We must however repeat the following points.
a) Wear **suitable protective clothing** (including safety glasses or goggles) whenever refrigerants are handled, or refrigeration systems opened to atmosphere
b) Always **purge refrigerants to the outside atmosphere** when a system is to be emptied
c) **Do not braze or weld systems containing refrigerants or refrigeration oil**
d) **Do not blow through systems which might contain oil, with oxygen or other potentially explosive gases**
e) Use only **suitable materials**, and **good practices**, when working on any type of system.

Adjustment of controls

A number of adjustments do not require circuits to be opened – especially if systems have adequate provision for instrumentation. We will review them before passing on to more complex work, involving opening refrigeration systems.

Thermostatic expansion valves

Before adjusting the superheat screw of a TEV, because an evaporator is receiving too much or too little refrigerant, make sure that the sensing bulb is firmly connected to the suction line and is correctly positioned (see Chapter 6), and that neither the bulb nor the capillary has been damaged and part or all of the bulb charge lost. Check also that all capillary tubes from the refrigerant distributor (if fitted) are

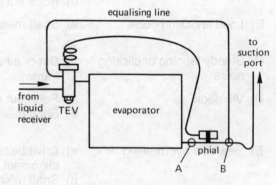

Fig. 11.1 TEV superheat test positions

undamaged, and passing refrigerant.

If the above checks prove satisfactory, check the valve superheat as follows. Refer to Fig. 11.1.
a) Using (preferably) an electronic thermometer, or alternatively a sensitive dial thermometer with a gas-filled bulb, clamp the thermocouple or bulb securely to the suction line at point A.
b) Using a self-sealing access valve at point B, fit a gauge to read the evaporator outlet pressure and, using a pressure/temperature chart, convert this to the equivalent temperature. (N.B. suction temperature can be calculated from the pressure at the compressor suction port, minus say 1 °C or a more accurate estimate of pressure drop in

the suction line; but this is *not* as accurate as the method first described.)

c) Subtract temperature B from temperature A to determine the superheat of the system. For example:

Temperature at A ....5 °C (41 °F)
Pressure at B = 62 kPa
(9 psig) which if using
R12 is equivalent to ....0 °C (32 °F)
―――――
Therefore superheat .... 5 °C (9 °F)
═════

Too low a superheat means that too much refrigerant is entering the cooling coil. The suction pressure is high, and the suction line may frost back to the compressor.

Too high a superheat means that the evaporator is starved. The suction pressure is low, and only part of the cooling coil is fully used.

To adjust superheat, remove the cap from the adjustment screw, and turn it not more than one quarter of a revolution at a time. Note the effect on the temperature at point A (remembering that it can take up to half an hour for a valve to settle down again). Unfortunately, TEV manufacturers have not standardised settings – one cannot, for example, say 'turn anti-clockwise to reduce superheat'. To be safe, look up the technical data on the valve concerned, or be prepared to spend time getting the correct setting by trial and error.

If it is necessary to strip a TEV for any reason, the system should be pumped down and valved off so that the valve is at a slight positive refrigerant pressure – say 14 kPa (2 psi). This system pressure will resist the entry of air or other contaminants.

Capillary tubes

Instructions of questionable value appear in some service manuals, in the form of advice on clearing blockages from capillary tube refrigerant controls. The procedures – normally involving the use of a hand pump, and hydraulic oil to force the contaminant out of the capillary – are usually effective. But the cost of the operation, in terms of materials and labour, often exceeds that of a shiny new capillary control of guaranteed quality!

If a capillary tube is blocked, remove and replace both it *and* the liquid-line strainer. If it is necessary to change evaporator or condenser, replace both capillary and liquid-line strainer.

Solenoid valves

The most frequent cause of trouble with solenoid valves – be they liquid-line, hot gas bypass, or reverse-cycle/defrost controls are:
a) dirt, sludge or other system contaminants; or
b) valve bodies distorted by excessive heat when brazed or soldered into place.

Either may result in any of the common symptoms 'valve will not open'; 'valve will not close'; 'valve is not seating properly' or 'the coil has burned out'.

Cross-sections through various types of solenoid valve appear in Fig. 5.21 (reversing) and Figs 6.24/25 (liquid-line). Figure 11.2 is an exploded view of a liquid-line valve, which will help you to work

Fig. 11.2 Liquid-line solenoid valves

out how to strip one, and how to reassemble it.

In the event of trouble with a solenoid valve, pump down to a slight positive pressure and check it for wax, carbonised oil or other contaminants. If any are present, strip and thoroughly clean the valve and **renew the refrigerant strainer ahead of the valve**, as well as the strainer in the entry port of the valve (if applicable).

Distorted valve bodies can only be replaced. If doing this, note the procedures detailed in Chapter 8 and point the flame away from the valve body. If possible, favour designs with longer, not shorter sweat connection 'tails'. Valves *must* be installed in the same plane (horizontal or vertical) as the original, and be of the same make and type – some will only work if plungers are vertical (i.e. the body is horizontal) but others have strong enough coils to work in any plane.

Hot gas bypass valves

These capacity regulating valves are installed in a bypass between the discharge line from the compressor and (ideally) an entry port between the expansion valve and distributor. They modulate towards their fully open position as suction pressure falls. Figure 11.3 shows a typical installation layout.

The set point pressure is adjusted by the control spring, and the valve remains closed so long as evaporator pressure is above the preset level. If it falls below the set point, the valve starts to open and hot gas is injected into the entrance to the evaporator. This increases suction pressure, and the load on the compressor,

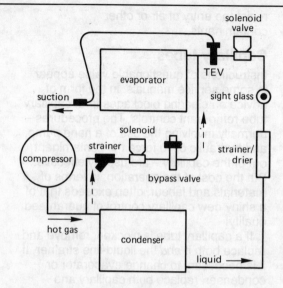

Fig. 11.3 Hot gas bypass line

and prevents frosting back or liquid slugging. Note from Fig. 11.3 that a solenoid valve is installed in the discharge line ahead of the bypass valve (it is wired in parallel with the liquid-line solenoid valve) to prevent the refrigerant pump-down system from being bypassed and rendered inoperative.

Figure 11.4 is a cross-section of a typical hot gas valve. To adjust such a valve, fit a suction gauge and coarse-set the desired suction line pressure by adjusting the liquid-line valve. Finer settings can then be made by retensioning the adjustment spring.

Bypass valve problems are not common if systems are free from contaminants. *Two* types of defects can be found.

a) Valve will not open – due to the seat or piston being affected by sludge or other contaminants. Pump down to a slight positive pressure, and strip and clean the valve
b) Valve will not close – due to one of:
  i) presence of contaminants – see (a) above
  ii) bypass tube blockage or restriction – clean or replace bypass, and strip and clean valve
  iii) diaphragm failure – replace valve power element.

Head pressure controllers

Two main types of controller are available:
a) pressure operated – actuating dampers at one or more evaporator fan outlets *or* electronic motor speed controllers;
b) temperature operated – actuating electronic motor speed controllers *or* cycling one or more fan motors on/off in accordance with ambient conditions.

The second type of controller may be less effective than the first, since there is not necessarily a fixed relationship between system pressures and ambient temperatures.

Controllers should be set to be fully effective (i.e. motor at lowest speed, or dampers fully closed) at a liquid-line pressure slightly above the minimum level nominated by the manufacturer of the compressor and then operated to increase condensing efficiency as refrigerant pressures/temperatures increase. A typical 'as fitted' circuit is as shown in Fig.11.5.

Head pressure controllers can only be re-set using specific instructions issued by their manufacturers.

REFRIGERATION SYSTEM SERVICE

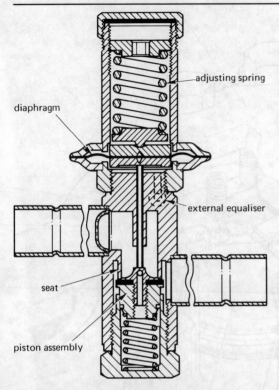

Fig. 11.4 Bypass valve cross-section

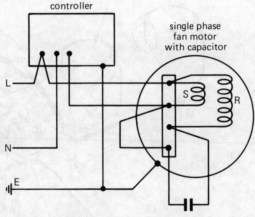

Fig. 11.5 Head pressure controller layout

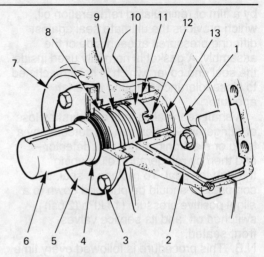

1. Front main bearing housing
2. Oil drain to crankcase
3. Cover setscrew
4. Stationary seal ring
5. Gland cover
6. Crankshaft
7. Front cover
8. Rotating seal ring
9. Spring
10. Spring sleeve
11. 'O' ring
12. Oil feed to gland housing
13. Main bearing

Fig. 11.6 Compressor shaft seal section

condenser with a sheet of hardboard, and note the pressure at which the control operates (or fails to operate!). For a low pressure control, system pressures can be steadily reduced by throttling down at the liquid-line valve. In either case, be ready to remove the 'obstruction' quickly if pressure changes are greater, and occur more quickly, than expected.

Pressure controls

A typical pressure control section is illustrated in Fig. 6.7. To adjust control settings, fit suction and discharge pressure gauges and make only small changes to settings before checking effects on system procedures.

Control operation, and the accuracy of existing settings, can be tested by 'rigging' the equipment. For a high pressure control, progressively blank off an air-cooled

Accessible and open compressors

Compressor shaft seals

The weakest point of an open compressor is its shaft seal, which is likely to leak oil and refrigerant as the result of mechanical damage by misaligned drive kits (see Figs 8.37, 38, 42 and 43) as well as 'fair wear and tear'.

A typical shaft seal is illustrated in Fig. 11.6. The seal is contained between a step (or 'shoulder') in the shaft, and a bolt-on seal (or 'gland') cover. The shaft passes through the seal and cover, to carry the half shaft or vee belt pulley.

The wearing surfaces of a seal are held in place by a spring, or bellows. This is slipped over the shaft, and secured at the shaft 'shoulder' by a sleeve. At its front end, the spring presses a carbon sealing ring against a matching metallic surface. The two wearing surfaces are lapped to optical standards of flatness and covered

191

TROPICAL REFRIGERATION AND AIR-CONDITIONING

by a film of recirculated refrigeration oil, which provides the ultimate seal against differing pressures at each side of the assembly. A gasket is normally used inside the seal end cover, which is itself machined to close tolerances around the shaft aperture.

A shaft seal leak is indicated first by loss of refrigerant – which can be detected by a lamp or electronic refrigerant detector – and then by refrigeration oil seeping through the seal. To replace a seal, the compressor should be pumped down to a slight positive pressure (14 kPa (2 psi)), switched off, and its service valves front-seated.

N.B. This procedure is followed every time it is necessary to open an open or accessible hermetic sealed compressor or system component.

The seal cover is then carefully unbolted, leaving two bolts at opposite extremes of the cover until last, so that any excess pressure can be released in a controlled fashion. When the cover has been removed, the seal components can be removed by hand, or using a seal withdrawal tool.

The wearing surfaces of seals which are leaking but not badly damaged can be repaired by carefully relapping.

Compressor valve plates

The valve plate and suction/discharge reed assemblies can be easily removed after an open or accessible hermetic compressor has been pumped down. Valve plates can be relapped if slightly worn, but there is little future in attempting to repair damaged valve reeds. Note that new gaskets should

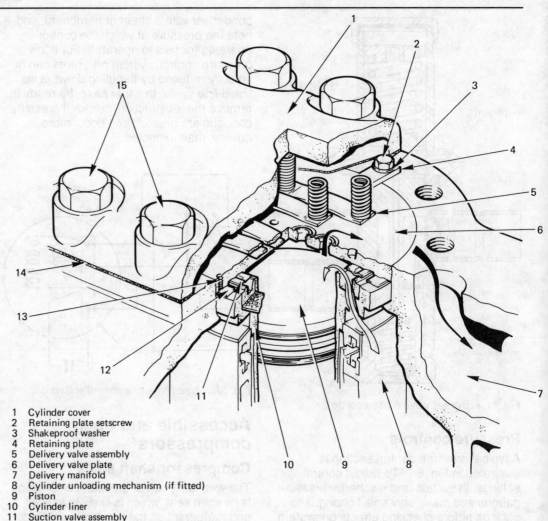

1 Cylinder cover
2 Retaining plate setscrew
3 Shakeproof washer
4 Retaining plate
5 Delivery valve assembly
6 Delivery valve plate
7 Delivery manifold
8 Cylinder unloading mechanism (if fitted)
9 Piston
10 Cylinder liner
11 Suction valve assembly
12 Metal joint (0.25 mm thick)
13 Jacking-off hole
14 Cover joint
15 Cover setscrews

Fig. 11.7 Valve plate assembly

REFRIGERATION SYSTEM SERVICE

always be fitted when a compressor component is removed for scrutiny or repair.

Stripping compressors

After systems are pumped down, and seal end covers removed, compressors can be separated from their service valves and the oil poured into suitable containers. Examine the oil to see if it contains any swarf or contaminants, but do *not* use it again in the compressor.

After removing the rear (or motor) end cover and gasket, the compressor pistons, connecting rods and crankshaft can be removed in that order (see Fig. 11.8). *Con rods* and pistons must be marked, to ensure that they are replaced in their original positions. Any damaged items must be replaced and all components should be stored in clean refrigeration oil until it is time for them to be cleaned (preferably using R11 or a non-toxic, non-explosive cleansing/grease solvent; caustic soda and trichloroethylene are *not* recommended) and replaced. Do not forget shaft or other bearings, which must be examined for fit and wear and must be replaced if their condition is less than excellent. Oil strainers and channels must also be carefully inspected, and any contaminants removed.

When re-assembling compressors, take care not to use the original gaskets (or other sealing materials) or oil. All components (including the crankcase and other castings) must be cleaned and examined to ensure they are free of defects, then coated with fresh, dry refrigeration oil.

Care must be taken not to over-tighten bolts, and the use of torque-indicating wrenches is desirable. We list a logical

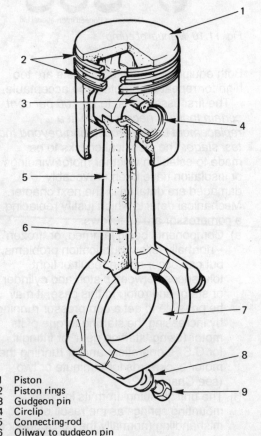

1 Piston
2 Piston rings
3 Gudgeon pin
4 Circlip
5 Connecting-rod
6 Oilway to gudgeon pin
7 Bigend cap
8 Washer
9 Setscrew

Fig. 11.8 Piston and con rod assembly

Fig. 11.9 Section of accessible hermetic compressor

order in which components can be refitted.

a) Crankshaft bearings (often grooved or marked to indicate which side should contact the thrust surfaces of the crankshaft).
b) Crankshaft (inserted from the front (seal) end) then the seal end cover.
c) Oil pump end bearing head. Adjust the end play of the crankshaft by shims between the bearing head and housing. Double check the end play with a feeler gauge after the bearing head is positioned and tightened.
d) Remove the seal end cover, and insert the shaft seal. Replace the end cover.
e) Install the oil pump and suction strainer.
f) Assemble and test (i) valve plates and reeds, and (ii) the cylinder unloader mechanism, if used.
g) Fit piston compression oil rings, if used.

TROPICAL REFRIGERATION AND AIR-CONDITIONING

h) Assemble pistons, con rods, and (if used) cylinder liners.
i) Properly position the crankshaft, insert pistons and con rods through cylinders, and complete the assembly to the crankshaft.
j) Refit the valve plate and cylinder head.
k) Replace the compressor on its base, and connect the service valves.
l) Insert a new charge of dry refrigeration oil.
m) Evacuate and leak test the compressor.
n) Return the unit to operation.

Adding oil to open/accessible hermetic compressors

Never pour refrigeration oil into a compressor crankcase – it is likely to be contaminated by air and moisture in the process.

With the compressor pumped down as instructed, remove the oil filler plug, and insert new, dehydrated refrigeration oil using a fully-primed hand pump (see Fig. 11.10). Make sure that air entering the can to replace oil used is passed through a drier and that both the compressor and the oil can are sealed immediately the correct quantity of oil has been added.

Replacing welded hermetic compressors

The replacement of a welded hermetic compressor after a motor burn-out is potentially the most critical operation we are likely to perform. Unless the job is planned and carried out to the highest standards, there is every possibility that the replacement compressor will also burn out.

Fig. 11.10 Manual of pump

Both equipment and labour costs are too high for repeat burn-outs to be acceptable.

The first essential is, to *be 100 per cent certain that it is necessary to fit a replacement before proceeding beyond the test stage*. The electrical checks to be made to establish whether motor windings or insulation have been irrevocably damaged are detailed in the next chapter. Mechanical defects which justify replacing a compressor are as follows.

a) Components being jammed, or 'frozen' – normally follows lubrication problems, but can occur as the result of tight tolerances between piston and cylinder or stator and rotor. In this case, it may be possible to get a compressor running by increasing the starting torque of its motor (using 'starting gear' or fitting it for C.S.R. operation) and/or running the motor backwards for a minute or two (see Chapter 12).
b) The unit 'jumping' from its internal mounting springs as the result of mishandling (normally the consequence of an appliance not standing upright during transit). The symptom is an abnormally high noise level (hammering), frequently followed by loss of pumping efficiency as the result of internal pipelines fracturing.
c) Damage to suction or discharge valves, or other components, resulting in unacceptably low pumping efficiency. This condition can be positively checked if the compressor does not incorporate gauge ports, by installing line-tap or Schrader valves (see page 196) close to suction and discharge tubing connections to the compressor casing.

The most serious damage is that which results from a compressor motor burning out. This causes oil to carbonise, form sludge, and contain acids as well as pieces of charred winding insulation and electrical materials. All traces of these contaminants must be removed from the system when the compressor is replaced, since their presence inevitably leads to a repeat burnout.

Burned out compressors must be handled with caution. If the seal of the service (or 'slave') tube is broken, the characteristic smell of burned windings is unmistakable and the compressor oil will be dark, evil-smelling and contaminated with carbonised particles. Contaminated refrigerant should be purged to fresh air, using a disposable length of tubing, before the compressor is removed; and protective clothing (goggles and gloves) must be worn to prevent refrigerant, oil or acids from injuring skin or the eyes.

Compressor replacement procedures are given below.

a) With **power off**, remove all electric conductors and accessories. Using a pipe cutter, cut suction and discharge lines close to (within 50 mm of) the compressor housing, unless service

valves are fitted. Release the external mountings and remove the burned-out compressor.
b) Remove the original capillary tube and refrigerant strainer.
c) Using a purpose-made pump, thoroughly flush out the remainder of the system with R11. Continue until no traces of oil or contaminants emerge from the suction line and/or suction-line accumulator.
d) Fit a new capillary tube, an over-sized liquid-line strainer/drier, the new compressor and a suction-line filter-drier (or 'burn-out'/'clean up' kit) immediately upstream of the compressor. If this does not have service valves, install Schrader valves before and after the suction-line drier. All electrical accessory items (capacitors, relay, overload etc.) *must* be replaced with new equipment.
e) Carefully silver solder all new joints whilst washing through the system with dry nitrogen applied at low pressure (see Chapter 8).
f) Evacuate the system, using either the Triple Evacuation Method at 710 mm (28 in) Hg or, preferably, a high vacuum pump recording 250 microns or better on an electronic vacuum gauge (see Chapter 8). **Under no circumstances use the compressor as a vacuum pump.**
g) Leak test, by making certain that gauge pressures do not increase (other than in step with room temperature changes) for three hours if using the Triple Evacuation Method, or one hour if using an electronic high vacuum kit.
h) Using a fully-equipped charging station, add the correct refrigerant (and if applicable oil) charge.
i) Remove the charging equipment, remake all electrical equipment and start up the new compressor.
j) Check the pressure drop across the suction-line filter-drier, using a differential pressure gauge if possible and check the amperage drawn by the compressor.
k) After one hour, recheck the pressure drop. Ideally, this should not exceed the values shown in Fig. 11.11.
l) After three hours, again check the pressure drop. If this exceeds three times the value quoted in Fig. 11.11, fit new liquid and suction-line driers. This procedure must be repeated until the pressure drop stabilises at or below the levels indicated above, for permanent operation.
m) Where the replacement compressor is a second or third replacement, i.e. the system is one in which there have been several burn-outs, an oil sample should be taken and acid-tested when the suction line drier pressure drop is at an acceptable level. Only if the oil is confirmed to be acid-free can the operation be regarded as satisfactorily completed.

The above procedures are fully current, many being based on Tecumseh recommendations. The sequence of operations and end objectives is of course the same in systems using welded or accessible hermetic compressors. The only changes are those resulting from variations in evacuation and leak testing procedures when working on large systems.

Gas charging methods

Conventional procedures are: to charge gas from a vertical cylinder, standing charging valve upwards; or to charge liquid from an inverted cylinder with the charging valve beneath the liquid refrigerant level. This is no longer an invariable rule, since du Pont now pack some Freon refrigerants in cylinders with a new type of charging valve. This enables liquid or gaseous refrigerant to be selected at the valve, with the cylinder remaining upright.

Air or overcharge?

Abnormally high condensing pressures can result both from an overcharge of

Evaporating Temperature	R12		R22		R502	
	kPa	(psi)	kPa	(psi)	kPa	(psi)
High (A/C)	13·9	(2)	20·7	(3)	20·7	(3)
Medium (Refrigeration)	10·3	(1½)	13·9	(2)	13·9	(2)
Low (Frozen Foods)	3·4	(½)	3·4	(½)	3·4	(½)

Fig. 11.11 Clean-up kit pressure drops

refrigerant and from air in the condenser. The question 'which?' is not difficult to solve.

If a head pressure gauge is fitted and the system switched off, then:
a) if there is an overcharge of refrigerant, the condensing pressure will fall slowly but steadily until it is equivalent to room temperature.
b) if there is air in the condenser, pressure will remain well above that of the equivalent room temperature.

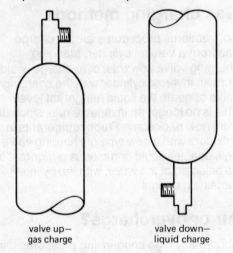

Fig. 11.12 Charging cylinder positions

Pressures in welded hermetic systems

The ready availability of line-piercing valves which can be easily and permanently installed in lines up to $\frac{7}{8}''$ OD has removed the major difficulty in diagnosing faults in small all-welded systems. They are not expensive, and their use can only help to avoid errors and repeat service calls (see Fig. 8.7).

When piping up larger, field-assembled systems the use of Schrader valves is desirable, even when compressor and liquid line valves are provided. This type of valve costs even less than line-tap models and has many uses. In split systems, refrigerant can be charged at either condensing or airhandling units. Suitably located valves make it quick and easy to adjust TEV superheat, or the settings of head pressure controllers or bypass valves, to take oil samples, or to check pressure at or through any part of the system.

Rotolock welded compressor valves

An increasing proportion of welded hermetic compressors incorporate suction and discharge valves, which again help to eliminate guesswork. Some other 'cans' incorporate provision for the installation of service valves by equipment manufacturers – or distributors – who consider that service costs and standards are more important than minimal first costs. The following notes will be of help to those fitting replacement compressors.

Some compressors have 'spuds' brazed into their casings. If screw-on caps and seals are removed from 'spuds', Rotolock service valves can be screwed onto the fittings. Alternatively, suction or discharge line terminals can be silver-soldered to the screw-on capping rings, using appropriate precautions against over-heating compressor casings.

A second type of fitting uses a steel adaptor, which fits over and is silver-soldered to stub tube connections emerging from the compressor casing. (Don't forget to cut the end seals from stubs before fixing the adaptors!) The other end of the adaptor is threaded to receive a screw-on Rotolock valve (which has a capped service or charging port).

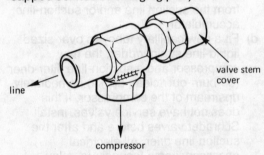

Fig. 11.13 Rotolock valve adaptor

The availability of new types of valves and test equipment will do much to improve standards of diagnosis, and service work. We advise students to make as much use of them as possible; there is no place for guesswork in modern technology, and a moistened forefinger can no longer be regarded as an acceptable instrument for system analysis.

Finally, the use of instruments to record the temperature and/or RH of the conditioned space, ambient conditions, chilled and process water temperatures, power supply voltages, compressor cycling times and other data over protracted periods of time is essential if it is necessary to obtain a complete and balanced picture of plant performance, without employing teams of personnel to keep day and night records.

Electrically or mechanically operated recording instruments covering a variety of applications are available – some forming an integral part of control systems. Figure 11.14 illustrates a two-pen, 24-hour DB and WB temperature recorder which is typical of a wide range of instruments available.

Workshop layout

The W.A.E.C. syllabus includes workshop layout requirements. We do not propose to discuss this subject in detail, since the most effective layout will vary with the type and number of items of equipment being processed; but a few general 'ground rules' can be established.

Within the limitations imposed by the shape and size of building available, internal layouts should enable a piece of equipment such as a domestic refrigerator or a room air-conditioner to follow a logical sequence.

a) it enters a reception room, where defects can be analysed and service requirements noted on job cards.
b) If necessary, it can be side-tracked through a cleansing bay where heat exchangers and other metallic components are cleaned and reconditioned and ferrous items repainted with a corrosion inhibitor. If a spray booth is needed, it should be located in this general area.
c) It passes into a clean, dry and well-ventilated service shop equipped with mobile bases for refrigerators, or trolleys with component shelves for room units. Where the number of units justify it, install multiple power points and evacuation/charging/leak testing stations.
d) After being serviced in accordance with the instructions on the job card, it is sent to an 'outgoing test bay' to be finally tested for performance and work quality before being returned to owners.

The service shop will of course require an adjacent parts/tools/test equipment store and it is sometimes necessary to have a separate locked storage area for equipment awaiting the arrival of special components, or the acceptance (or otherwise) of quotations for repairs. In any case, all appliances must move in one general direction and the checks made before they are allowed out must be as comprehensive as those made when they are received. Equipment requirements have been indicated in notes on specific procedures, but one additional item has been saved until last: adequate protective covers in which reconditioned appliances can be redelivered to their owners without suffering dents or scratches during transit. An obvious need, but one which is too often ignored!

Revision questions

1. The plant for a commercial cold room includes a 0·5 kW semi-hermetic compressor charged with R12. It is suspected that air has entered the system, and service pressure gauges have been installed.
   a) List two major observations made using the gauges which would confirm the presence of air.
   b) Enumerate two other ways in which the presence of air could be confirmed.
   c) Describe in detail the method used to eliminate the air.
   d) What signs would confirm that air has been present, the compressor valve cover and valve plate having been removed for inspection?
   (WAEC)

2. a) Make a labelled line sketch showing the layout of a commercial refrigeration evaporator and an externally equalised thermostatic expansion valve.
   b) With the help of your diagram, explain briefly how the TEV superheat setting can be checked and altered.

3. A compressor has been stripped and is to be reassembled. Give the order in which

Fig. 11.14 DB/WB temperature recorder

you would refit the following components:
oil pump end bearing,
shaft seal,
piston compression oil rings,
crankshaft,
cylinder valves,
oil pump,
valve plate,
crankshaft bearings,
seal end cover,
suction strainer,
piston and con rods.

12. Electrical service procedures

Safety precautions

Before adjusting or removing/replacing any electrical component, ensure that power and control circuits are isolated at a point in full view of the engineer, and restored only when it is safe to do so. All tools and test equipment must be suitable for use at rated power supply characteristics.

Electrical components are responsible for something like one half of all service calls and it is likely that on many occasions the work which ensues does not include a full check of the entire system, and analysis of the cause of the original failure. For instance, a defective starting relay could be replaced, and no test made on the condition of the starting capacitor – which might have caused the relay to fail in the first place. It is also possible that an appreciable percentage of compressors which are considered to have 'burned out' are in fact perfectly sound – an internal thermal/current overload might for example have opened and produced the symptoms of an 'open circuit' winding, but would remake contact if given sufficient time to cool down. It is equally possible for an internal overload of the pressure-operated type to open, and for the compressor to be labelled 'not pumping' by someone unaware of the presence of the overload device.

We therefore recommend not only that *all* components in an electric circuit are tested if one proves defective, but that the best available test equipment is used before compressor replacement is authorised. Expense prohibits the issue of sophisticated equipment to every service man, but all adequately equipped workshops should contain the equipment needed to carry out each test referred to in this chapter.

Test equipment

Circuit tester

A two-probe tester with insulated handles and a neon indicator is essential – never touch a circuit component with a hand or finger until you are *sure* that it is not live. A current in the control circuit is frequently an unexpected hazard which is not always removed by switching off power at an isolator serving one motor or starter. Always make the first contact with a circuit tester – which can effectively be combined with a voltage indicator.

A test cord can also be used to bridge thermostats, overloads, etc. to determine whether or not they are making circuit. First ensure that it is capable of carrying enough current.

Tic tracer

This is a useful instrument, which deserves to be better known and more widely used. If the sensor is run along a conductor,

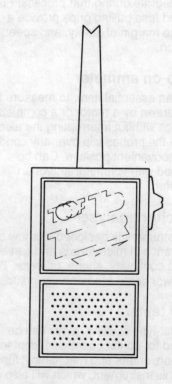

Fig. 12.1 Tic tracer

there is an audible confirmation if a current is flowing through it or silence if there is not. A tic tracer therefore enables quick and positive checks to be made on fuses forming part of large fuse boards, or on the condition of cables if there is power at a socket, but not at motor terminals.

Fuse puller

Ceramic fuse holders are generally withdrawn by hand – and have been known to disintegrate during that process! Fully insulated fuse pulling grips provide a welcome margin of safety, and speed the operation.

Clamp-on ammeter

This is an essential item, to measure the power drawn by a motor or a complete appliance without interrupting the electric circuit – the probes clip over any conductor in any convenient position. Can be combined with voltmeter and/or an ohmmeter.

Capacitor analyser

This instrument is designed to show actual capacitance in microfarads and can also indicate continuity, shorts, or leaks. Test and power cords are normally included.

Relay tester

Potential-type starting relays can be checked for contact condition, contact operation, shorts and mechanical function using this instrument, which will also give accurate readings of operating (*pull-in* and *drop-out*) currents.

Fig. 12.2 Clamp-on ammeter

Fig. 12.3 Battery-type ohmeter

Battery-type ohmmeter

This is used to test the resistance and continuity of hermetic unit motor windings, the scale coverage being from 0·25 + to 200 ohms. It can also be used to check the continuity of hermetic unit motor windings, coils – the tests on relays described later in this chapter could, for example, be made using a battery-type ohmmeter.

'Meggers'

Hand-cranked megohmeters are the conventional means of testing motor or other windings, or controls, the test current being generated when the handle is 'wound up'. Meters indicate continuity in ohms and insulation values in megohms (1 × 10^6 ohms).

High potential tester

This instrument provides the most conclusive tests on motors. Insulation breakdowns, leaks, shorts or grounds are all indicated by a lamp and/or a buzzer.

Fig. 12.4 'Hi-Pot' tester

The test voltage should be approximately (2 × appliance nominal volts) + 1 000.

A 230 volt motor should, for example, be tested with a 'Hi Pot' tester set for 1 450 volts.

Hermetic unit analyser

This is the most complex, and potentially the most valuable item of electrical test equipment designed specifically for the refrigeration engineer. A typical unit analyser contains both ammeter and voltmeter and circuitry to test or perform the following functions on single-phase compressors up to 7·5 kW (10 hp) nominal capacity.

a) Indicate the line voltage.
b) Indicate starting/running amperages.
c) Confirm the continuity of motor windings, and the insulation value.
d) Provide the capacitance required to start a compressor not wired into its appliance circuit
e) Reverse the direction of motor rotation to try to free jammed ('frozen') mechanical components, by means of a rocker switch.

Test sequences

It is necessary to establish and keep to a set and logical order of progression when testing the individual electrical components of an appliance.

The simple wiring diagrams for domestic refrigerators (Fig. 5.11) and air-conditioners (Fig. 5.19) will help in preparing schedules. The domestic refrigerator will, for example, be seen to have two series of controls bridging the line and neutral conductors. One serves only the cabinet light, actuated by a door-operated micro switch. The second links the thermostat (and defrost heater if fitted), thermal overload, compressor and starting relay. If a machine will not start when switched on, the first check should be that there is power at the fused switch. If there is, the logical progression should be in the order already quoted – there is no point in checking the compressor before the thermostat.

This type of block layout analysis, and order of component testing, should be applied to all electrical service. The logic will vary with different types of appliance – for example, the room air-conditioner has no light, but does introduce an operating mode selector and fan speed switches, and a fan motor with – usually – P.S.C. wiring, and the associated capacitor. The notes which follow do not attempt to follow a system layout, but group components by type and run from the more simple to the more complicated procedures, to enable students to follow the 'building block' methods of progression used when looking at equipment design practices. Space does not allow us to quote detailed test procedures for each type of single-phase motor and their accessories; it is the responsibility of appliance manufacturers to provide adequate wiring diagrams, to identify clearly all components used, and to quote permissable tolerances where applicable.

Starting relays

The following types of starting relay are in common use.

Current relays
These are described in Chapter 5 and illustrated in Fig. 5.13. The basic arrangement of contacts etc. is shown in Fig. 12.6. Whether used with R.S.I.R. or C.S.I.R. motors, starting contacts I and S **are normally open**. The continuity between these contacts should be tested on the premise that if there *is* continuity between them, the contacts have stuck or been 'welded' together, and the relay will

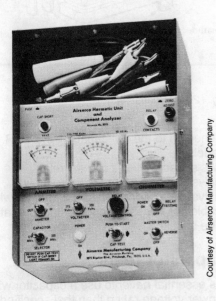

Fig. 12.5 Hermetic unit analyser

TROPICAL REFRIGERATION AND AIR-CONDITIONING

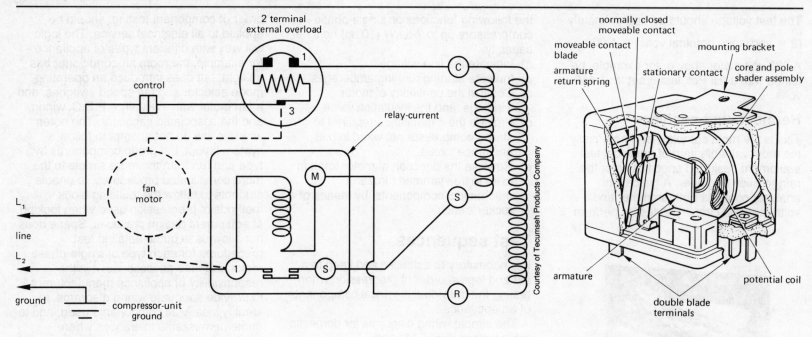

Fig. 12.6 Current relay contacts

Fig. 12.7 Potential relay

probably have to be replaced.

Secondly, it is necessary to test the continuity of the current relay coil between I and M. If there is no continuity, the coil is open circuit and either it or the relay must be replaced.

Potential relays
These operate in a different fashion, with high starting torque C.S.I.R. and C.S.R. motors, and also P.S.C. motors which have been fitted with starting gear in the field. (P.S.C. motors do not normally use starting relays or starting capacitors.)

In this type, the **contacts are normally closed**. The coil senses increased voltage in the start winding as the motor picks up speed. When the voltage reaches a pre-determined level, the armature opens the contacts and de-energises the start winding. Power to keep the contacts open whilst the motor is running is induced in the start winding. If the motor is switched off the contacts close when the coil is de-energised.

Referring to the contact sketch in Fig. 12.8 the relay is proved defective if there is no continuity between 1 and 2 **or** between 2 and 5. Note that terminal 3 is a dummy with C.S.I.R. wiring, but is connected to the run capacitor in C.S.R. models.

As stated earlier, potential relay analysers can be used to thoroughly analyse relay condition, but a Megger will enable coil continuity and contact operation to be checked.

Capacitor tests

It is essential *never* to use a capacitor with a lower voltage rating than that specified. Higher voltage ratings are however acceptable.

The condition of capacitors is most

ELECTRICAL SERVICE PROCEDURES

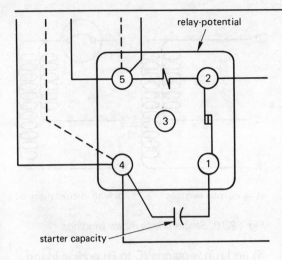

Fig. 12.8 Potential relay contacts

important. If they discharge across the contacts of starting relays, contacts can be fused and the relays rendered useless. Faulty capacitors can also damage motor start windings, and lead to burn-outs.

P.S.C. motor start capacitors are best fitted with bleeder resistors wired across their terminals. If it is necessary to do this in the field, a two watt, 15 000 ohm resistor can be soldered between the capacitor terminals.

Start capacitor values are specified by compressor or appliance manufacturers. Excessive capacitance increases both starting current and the temperature of start windings, and may reduce starting torque.

Run capacitors have a tolerance of only plus or minus 10 per cent of specified capacitances. To be safe:
a) do not use capacitors with *less* than specified values,
b) do not exceed specified values by more than:

specified values	maximum excess
10–20 μF	2·5 μF
20–50 μF	5 μF
50 and over	10 μF

(μF is the symbol for microfarad.)

Efficiency tests

A number of appliance manufacturers recommend testing capacitors in a circuit similar to that shown in Fig. 12.9. Procedures are:
a) remove the terminal connections
b) discharge the capacitor with a screwdriver
c) check that the resistance of the insulation between the terminals and the metal casing is not less than 50 megohms

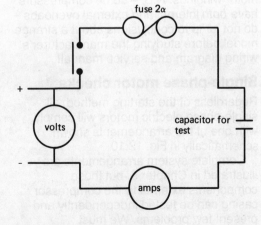

Fig. 12.9 Capacitor test rig

d) wire the capacitor into test rig as shown
e) switch on, note the voltage and amperage, and switch off quickly
f) check that the amperage is within 10 per cent of the ratings quoted for each size of capacitor at each of several voltage levels, on a chart or tables provided by the manufacturer.

However, the above procedures take a good deal of time, cannot easily be followed on site, and do not give a precise picture of the condition and performance of the capacitor. If a capacitor analyser similar to that illustrated in Fig. 12.3 is used, the following tests can be run without removing the capacitor from the appliance.

a) With the power off, disconnect the capacitor from all external wiring (including bleed resistors) and discharge it.
b) Connect the coloured analyser leads to the capacitor terminals in accordance with operating instructions; plug in and switch on the analyser.
c) Select 0–500 μF range, switch on and note the reaction of the light:
   i) the lamp lights, then goes out, (perhaps after a few minutes) – capacitor taking charge,
   ii) the lamp does not light – capacitor open circuit,
   iii) the light comes on and stays on – capacitor has shorted,
   iv) the light comes on, but then dims – capacitor is leaking.
d) If the capacitor works as in (i), press the button for the actual reading of the capacitance. (If necessary change from the 0–500 to the 0–100 μF scale.) Discard capacitors with capacitances

203

outside the specified band of tolerances.

This sequence is typical – the characteristics and operating instructions of different models vary – and shows that the use of this type of equipment will provide accurate information quickly.

Compressor motor tests

External indications

One can pick up useful information of possible faults before checking the windings of hermetic compressor motors. Problems likely to prevent their starting include the following.

Low voltage
This reduces starting and running torque; at 90 per cent of rated voltage, a motor has only 80 per cent of its design starting capacity. Ten per cent is the maximum acceptable voltage reduction. Three external indications of low voltage are given here.
 a) The motor hums, does not start or trips on overload.
 b) Lights dim when the motor is energised.
 c) Television pictures suffer when the motor is energised

If any of these symptoms are found, check the voltage first!

High pressure
Most systems have features – such as capillary tube refrigerant controls, or pressure equalising ports in expansion valves – which enable head and suction pressures to equalise, wholly or in part, during off-cycles. The maximum acceptable start-up pressure for P.S.C. motors used in room air-conditioners is 1 150 kPa (170 psig). External symptoms include overheated compressors cycling on safety devices.

Tripping fuses
Tripping fuses or circuit breakers *can* be undersized! Correct values should be, not less than 175 per cent and not more than 225 per cent of full load amps (FLA). Check their capacity if they have tripped.

Run capacitors
These can prevent P.S.C. compressors from starting, and repeated tripping usually causes overload devices to operate. Internal overloads can take more than an hour to reset – the compressor casing must cool to approximately 55 °C (130 °F) before this can happen. Until then, the system will present every symptom of open circuit motor windings. N.B. – some compressors have *both* internal *and* external overloads – do not jump to conclusions about a strange model before studying the manufacturer's wiring diagram and service manual!

Single-phase motor checks

Regardless of the starting method, all single-phase electric motors will comply with one of the arrangements shown schematically in Fig. 12.10.

Complete system arrangements are illustrated in Chapter 5, but those components external to the compressor casing can be tested independently and present few problems. We must concentrate now on the start windings (C to

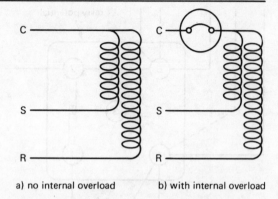

a) no internal overload   b) with internal overload

Fig. 12.10 Single-phase motor windings

S) and run windings (C to R) ending in the terminal box. Reference has already been made to the construction and characteristics of the internal overloads shown schematically in Fig. 12.10(b), and to methods of testing capacitors and starting relays, which will have been checked out before motor windings are confirmed to be faulty. At that stage, we have three alternative procedures from which to choose – using a megger, a high potential (Hi-Pot) tester or a hermetic unit analyser.

Using a megger
a) Referring to Fig. 12.10 or (preferably) the manufacturer's wiring diagram, check for continuity in the start winding (C to S) and run winding (C to R). If either winding shows *no* continuity, the compressor should be rechecked, and replaced if open circuit is confirmed.
b) Check for continuity between the common terminal (C) and a good

ELECTRICAL SERVICE PROCEDURES

earthing connection. If there *is* continuity, the unit is shorting to earth and must be replaced.

c) If there is continuity in both start and run windings, but it is difficult to start the motor, check the resistances of windings against manufacturer's performance data. For example, resistances through a typical Tecumseh hermetic unit windings should be (on 230V 60Hz supply) within the following tolerances:

Model	Resistance in ohms Start Wind	Both Windings
AJ5518E	3–7	4–8
AH5520E	2–5	3–7

Start wind resistances are measured between terminals C and S, and 'both' winding resistances between R and S.

It is usual practice for compressor manufacturers to examine compressors returned under guarantee; and there are obvious advantages in having megger readings rechecked if they are unfavourable, and before a compressor is removed from the circuit, using part of the standard warranty test procedure which follows.

Using a high potential tester

a) With the Hi-Pot tester set for 1 400–1 500 volts, and all mains power connections removed from the compressor terminals, (following the instrument manufacturer's specific operating instructions) connect one of the Hi-Pot leads to a good earthing connection, and the other to each motor terminal in turn, noting leakage values.

b) If there is no leakage, clip one Hi-Pot terminal to any one of the motor terminals, and the other to each of the remaining terminals, in turn. If there *is* leakage, it indicates continuity in the winding concerned.

c) If the results of tests made with Hi-Pot testers are favourable, try to start the motor, on mains power (the voltage should be tested first!). If the unit does not start, measure the voltage across terminals C and R when the motor is trying to start (locked rotor voltage). This should be above 90 per cent of the line voltage – for example, $230 \times 0.9 = 207$. If it is, the motor starting components should be rechecked before the compressor is written off.

The use of a Hi-Pot tester in the above cases provides a more searching test than a hand-wound megger. It is interesting to note the attempt in procedure (c) to start the motor, and run one last check on such components as capacitors and relays. These would not help if the compressor was mechanically stuck. The possibility of dealing with that situation, in a final attempt not to change a compressor unless it is absolutely necessary, is our third alternative.

Using a hermetic unit analyser

These instruments must be used strictly in accordance with the instructions issued by their manufacturers. They have similar functions which can be briefly summarised:

i) to check motor windings for continuity, open circuits, shorts or grounds

ii) to identify compressor terminals (S, R and C) when they are not known

iii) to start compressors without using any of the controls in the appliance circuit external to the compressor motor

iv) to attempt to start jammed or 'frozen' compressors by reversing the direction of motor rotation

v) to check the continuity and resistances of electrical equipment outside the motor.

The following procedures are intended to summarise typical test sequences followed when using a hermetic analyser with a single-phase compressor, and *not* to constitute specific instructions applicable to all analysers and all appliances.

a) Remove mains power connections from the appliance and with the analyser also isolated connect it to the compressor terminals, and to earth, using the conductor colour code identification data in owner's manual. Test for grounds by measuring resistance from each terminal to earth. The meter does not show any movement (i.e. resistance is infinite) if there are no shorts. Shorts would require the motor to be replaced. Repeat resistance tests between motor terminals (see *Using a megger* (a) above).

b) If the windings appear to be good, but the motor will not start: connect the analyser to mains power; select the appropriate control setting for the characteristics of the motor (including type(s) of capacitors used, if any); connect the leads in accordance with the owner's manual and follow the starting procedures for the unit. These will use capacitance provided by the

TROPICAL REFRIGERATION AND AIR-CONDITIONING

analyser, not the capacitors in the appliance. Energise the unit in three-second bursts. If the motor starts, check the appliance capacitors, relay etc.

c) If the motor does not start, the unit may be jammed. Rock the master switch several times between 'on' and 'reverse' positions, to attempt to overcome mechanical seizure by reversing the direction of movement of components. If the motor then starts, allow it to run for two minutes; then revert to usual starting procedures outlined in the last paragraph.

It will be seen that the use of this type of instrument enables usual megger tests to be carried out; it enables the user to attempt to start the motor without using any system controls, which might be faulty and it makes it possible to try to overcome seizures resulting from tight component fits.

Compressor motor fault analysis flow-chart

The sequence of tests to explore every possible defect in the electric system of a refrigerator or an air-conditioner with a single-phase hermetic compressor have been presented in the form of a 'flow chart' reproduced as Fig. 12.11. This is based on the use of a unit analyser, but provides a useful guide to the logic of testing motors with less versatile instruments.

Three-phase motor tests

Tests to be made on three-phase motors are in many respects more simple than

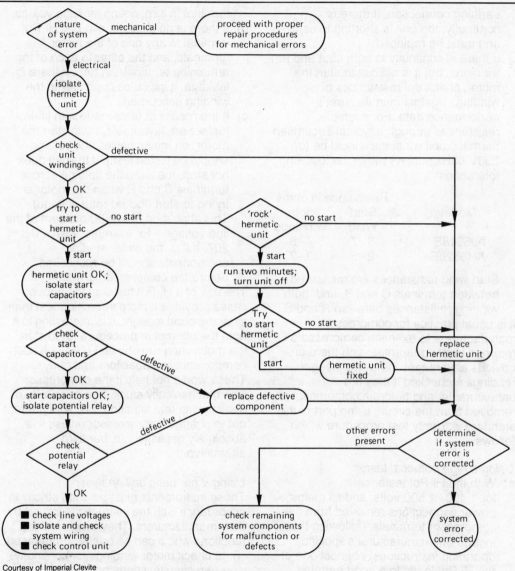

Courtesy of Imperial Clevite

Fig. 12.11 Electrical fault analyser chart

ELECTRICAL SERVICE PROCEDURES

those detailed for single-phase units. There are more windings to check, and the use of internally-mounted thermal type overloads is common. Externally, however, starting relays and capacitors are not required, a contactor or starter being used.

Hermetic compressor motors up to 0·55 kW (7·5 hp) nominal are usually started across line by simple contactors. Larger motors normally require the use of reduced voltage starters, which can take many forms – star-delta, part wind, and auto-transformer being most popular in this industry.

A simple, across line starter was illustrated schematically in Fig. 3.22 and it will be seen that, whether it is for manual or automatic operation, any operating defects can be traced using an ohmmeter.

Three-phase motor starters
We need not repeat the operating principles of various types of reduced voltage starters, described in Chapter 3. The service engineer may however wish to be reminded of *approximate* currents drawn at start-up by different types:

across line .... full load current × 6
star-delta .... full load current × 2
part wind .... full load current × 4
auto-transformer .... full load current × 2

These values vary with the torque against which the motor has to start, which in turn depends upon the degree of compressor unloading and system pressures. Starter construction, and the potential current reduction, are changed to suit individual compressor characteristics and there is no acceptable alternative to the use of manufacturers' wiring diagrams and performance data.

The wiring diagrams will also give details of electrical interlocks used in large systems to prevent motors starting in an unsuitable sequence. For example, water-cooled packaged water chillers are normally wired to ensure that both the chilled and condenser water pumps are running before the compressor is started. Do not ignore these interlocks, which are usually made via relays attached to individual motor starters – if line voltage control circuits are in use, the interlocks can be lethal unless they have been positively de-energised before service work commences.

Simple starters, and electrical interlocks, can be checked for continuity when isolated from mains supply, using either a megger or a battery-operated ohmmeter. The operation of more complex starters – such as the auto-transformer – and circuits with complex controls, electronic components etc. can be checked only by competent electricians who have been instructed in the details of the equipment concerned.

Other components

It is certain that students will, when working in the field, find it necessary to verify the condition of several other single-phase components, especially when working on room air-conditioners. Once more, there are variations between differing makes and models, but typical examples of the checks needed on some important components are described below.

Thermostats

The terminals of a typical air-conditioner thermostat are illustrated in Fig. 12.12. The contacts of the single-throw switch close on a rise in temperature and break on a temperature fall. The differential between make and break temperatures cannot normally be changed. The model shown has an auxiliary rotary switch to select: off; fan only; fan and compressor. Continuity tests should be made with an ohmmeter or megger, as follows:
a) at the maximum temperature setting continuity should exist between 1 and 6, 1 and 3, and A and B.
b) at the *off* position there should be no continuity between any terminals.
c) at the *fan only* position continuity is required between 1 and 3.

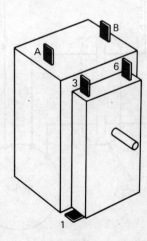

Fig. 12.12 Air-conditioner thermostat terminals

Multi-speed fan motors

Most room air-conditioners have three-speed, single-phase evaporator fan motors; a typical wiring schematic is shown in Fig. 12.13. Tests should prove that the motor will run under each of the following conditions:

a) with blue and black conductor terminals removed from the motor, and current applied through the brown (positive) to the white (neutral) conductor;
b) with the white conductor terminal removed from the capacitor and current applied through the brown (positive) to the blue (neutral) conductor;
c) with the blue conductor terminal removed, and the current applied through the brown (positive) to the black (neutral) conductor.

The capacitor can be independently checked, preferably using a capacitor tester. If the motor does not run under any one of the arrangements listed, it is almost certainly faulty and the state of its windings must be thoroughly checked.

Motor speed selector switches

The running speed of a fan motor is selected at either a rotary or a rocker switch, with *high, medium,* and *low* settings. A typical rocker switch terminal arrangement is illustrated in Fig. 12.14 and tests should confirm:

high setting – continuity between 2 and 4;
medium setting – continuity between 2 and 1, and 3 and 6;
low setting – continuity between 2 and 1, and 5 and 6.

Conclusion

In all cases, the essential requirements for control testing are to establish whether or not there is electrical continuity between contacts, and through coils and resistors. It is not recommended that mains voltages are applied – the current supplied by a megger or a battery-operated ohmmeter is preferred. When remote controls are used with packaged or central air-conditioners, it is usual for them to operate at low voltage (24 V) for the sake of safety. The control power transformer does not normally give trouble, but it is essential that it be checked for correct operation if external controls are inoperative. Similarly, low voltage controls are generally used in packaged water chillers and test procedures must ensure that they are not exposed to voltages which will burn out coils, conductors, etc. Always work from a manufacturer's wiring diagram and service manual, and test the components concerned in the manner which he details, so long as his procedures are logical. We feel that, to qualify for that description, tests must be as positive and informative as possible and we repeat our recommendation that work is carried out with the most modern and effective equipment available. The introduction of increasingly complex equipment signals the end of simple 'go or no go' procedures – only the best is good enough.

Fig. 12.13 Fan motor wiring

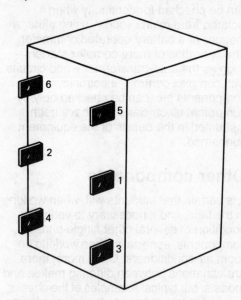

Fig. 12.14 Motor speed selector

ELECTRICAL SERVICE PROCEDURES

Revision questions

1. Make labelled line sketches to show the layouts of the electrical components and connections of:
   a) a current-type starting relay,
   b) a potential-type relay.

2. List the procedures to be followed when testing the main windings of a single-phase compressor motor in a welded hermetic compressor.

3. With reference to a three speed fan motor for use in a domestic air-conditioner:
   a) make a simple sketch of a typical motor speed controller and identify the electrical connections,
   b) state the main reasons for speed control of an *evaporator* and of a *condenser fan motor*.

13. First aid

Most accidents result from over-confidence, and carelessness and their effects are worse if people do not know how best to help someone who has been hurt. These notes are not intended as a course in medicine, but to help you to get priorities right, and act quickly and effectively, if something goes wrong nearby.

Get the priorities right. If someone is unconscious and touching or holding an electric cable or appliance, switch off the power before you touch him: do not risk electrocuting yourself. If there has been a fire, and you find someone badly burned with his clothes still burning, the first priority is to put out the flames and remove smouldering fabrics from the skin. In both cases, get help if you possibly can, so that first aid can be given whilst, not after, expert medical attention is being called.

Our notes review the types of injury – electrocution, burns, 'cold burns' and heavy bleeding – you are most likely to meet.

Electric shock

a) Switch off power. Make sure nobody can turn it on again (withdraw fuses etc.).
b) Call for help – as soon as possible, get someone to phone for a doctor, ambulance or the police. Remember that police officers are trained to take over in cases of emergency.
c) See if the victim is breathing, and if his pulse is beating.
d) If he is not breathing, but pulse is beating do the following.

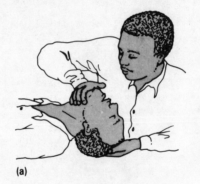
(a)

(c)

Fig. 13.1 Artificial respiration

i) Clear out the mouth (make sure tongue is not blocking air passage – remove any dentures, vomit, or other obstructions to breathing).
ii) With casualty on his back, tip the head backwards with one hand, raise his jaws and open his mouth

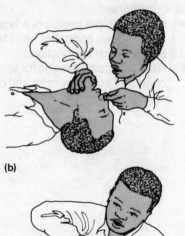
(b)

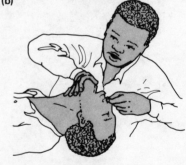
(d)

FIRST AID

with the other hand. Apply artificial respiration – seal nose with fingers, take deep and even breaths of air and blow it into the casualty's mouth – say eight times quickly, then at a steady ten per minute. Watch to see that his chest inflates when you blow in, deflates when you remove your mouth. If it does not, check the air passage again.

iii) Every minute, check to see whether casualty starts to breathe normally. If he does not, continue until medical help arrives.

e) If he is not breathing, no pulse, and his pupils are dilated follow this procedure.
   i) Give artificial respiration as at (d) above.
   ii) Apply heart massage – once a second, between oral breaths, lay both hands flat on his breast bone and press down gently but firmly to depress the chest wall 2 cm or so. Remove hands, quickly resume mouth to mouth treatment. Try to make both actions steady and consistent in pace and strength.

f) If the victim starts to breath steadily, turn him into the 'unconscious position'

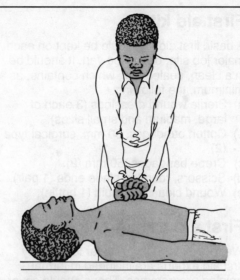

Fig. 13.2 Heart massage

– on the side with face turned down (head turned to one side). Pull up the top leg and arm, and pull up the chin to straighten the air passage. If possible, tilt stretcher etc. so that his head is about 300 mm below his feet. Keep him warm. Make sure that breathing and pulse continue until medical help arrives.

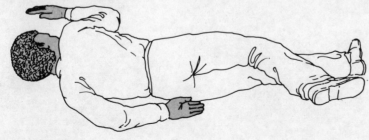

Fig. 13.3 The unconscious position

Heat burns

Heat burns must be cooled as quickly as possible. This reduces the severity of the burn, and helps reduce pain. Keep up a steady flow of clean water – the colder the better – for at least ten minutes.

Do not
a) apply lotions or grease;
b) touch or breathe on burned flesh;
c) move the casualty more than necessary;
d) cover burns with cotton wool or lint – clean sterilised dressings, sheets or pillow cases can be used to keep the wound sterile once the cooling process has been completed.

Do
a) remove any smouldering clothing – this is best cut loose, then washed off with a cold water spray;
b) get medical help as quickly as possible.

Cold burns

If the flesh or eyes have been damaged by refrigerant (or chemicals), **treat as for heat burns**. Irrigation with cool water will gradually raise the temperature of the flesh, and wash out any oil or chemicals. Continue until the casualty has feeling in the affected limb or area. Medical attention is essential.

Heavy bleeding

The loss of much more than a pint of blood can be fatal – heavy external bleeding must be stopped as quickly as possible.
a) Apply a pad of sterile material over the

cut or wound, and keep up gentle pressure.
b) If possible (i.e. if he has a cut, but no broken bones, in arm or leg) raise the injured limb above the level of the heart.
c) If the bleeding from a limb is arterial (the blood is bright red, and pumped out in steady jets) apply a tourniquet above the wound with a piece of material between the tourniquet and the skin. Note time applied, and release it as soon as possible without allowing heavy bleeding to restart. The tourniquet *must not* be left on for longer than fifteen minutes. The need for trained medical aid is urgent. Be sure that anyone who comes to help knows that a tourniquet has been applied.
d) Prevent infection – cover wounds with a sterile dressing or clean, non-lint cloth, as soon as bleeding is under control.

First aid kits

A basic first aid kit should be kept on each major job site, or service van. It should be in a clean, sealed box, which contains, as a minimum, the following.
a) Sterile wound dressings (3 each of large, medium and small sizes).
b) Cotton bandages, 50 mm, surgical type (2).
c) Crepe bandages, 50 mm (2).
d) Scissors, rounded blade ends (1 pair).
e) Wound cleansing liquid (1 bottle).

First aid training

We recommend that regular lessons by first aid experts are included in formal training programmes. These should cover artificial respiration, cardiac massage, methods of controlling heavy bleeding, the positioning of victims of accidents and the treatment of shock.

Fig. 13.4 A tourniquet

14. Conversion factors

The ability to convert from one to another of the various units of measure in current use is essential. The following *approximate* conversion factors are intended to help relationships to be established quickly. They need *not* be learned by heart, but the ability to instinctively relate one measure to another – for example, to know without calculating that 1 m^3 is about 35 ft^3 – is desirable.

Property	Units and conversion factors SI	Metric	British
Heat	kJ	kcal	Btu
	1	0·239	0·948
	4·187	1	3·968
	1·055	0·252	1
Specific heat capacity	kJ/kgK	kcal/kg °C	Btu/lb °F
	1	0·239	0·239
	4·187	1	1
Enthalpy	kJ/kg	kcal/kg	Btu/lb
	1	0·239	0·43
	4·187	1	1·8
	2·326	0·556	1
Latent heat	kJ/kg	kcal/kg	Btu/lb
	1	0·239	0·43
	4.187	1	1·8
	2·326	0·556	1
Length	m	m	ft
	1	1	3·28
	0·305	0·305	1
Area	m^2	m^2	ft^2
	1	1	10·76
	0·093	0·093	1

Property	Units and conversion factors SI	Metric	British
Volume	m³	m³	ft³
	1	1	35·3
	0·028	0·028	1
	1 litre = 0·001 m³		
	1 imperial gallon = 0·0045 m³		
	1 U.S. gallon = 0·0038 m³		
Mass	kg	kg	lb
	1	1	2·203
	0·454	0·454	1
	1 grain (British unit) = 0·065 g		
Density	kg/m³	kg/m³	lb/ft³
	1	1	0·062
	16·019	16·019	1
Force	N	kgf	lbf
	1	0·102	0·225
	9·797	1	0·203
	4·448	0·454	1
Pressure	kPa (1 000 N/m²)	kgf/m²	lbf/in² (psi)
	1	102	0·145
	0·0098	1	0·0014
	6·89	703	1
Power	kW	kW	hp
	1	1	1·341
	0·746	0·746	1
Performance	kJ/s	kcal/h	TR
	3·517	3 024	1
	1	860	0·284
	0·331	1 000	1·163
	1 TR = 12 000 Btu/h		

CONVERSION FACTORS

Property	Units and conversion factors		
	SI	Metric	British
Liquid flow rates	ltr/s	ltr/s	igpm
	1	1	13·2
	0·076	0·076	1
Heat transfer	$W/m^2.K$	$kcal/m^2.h.°C$	$Btu/ft^2.h.°F$
	1	0·860	0·176
	1·163	1	0·205
	5·678	4·882	1

Fig. 14.1 Conversion factors

Copper tubing dimensions

Copper tubing and fittings suitable for refrigeration duties are still manufactured and quoted in *inches* OD; and American tubing is made to several standards, for differing applications:

Type K – 'heavy' wall thickness, available in 'hard' and 'soft' grades. Suitable for refrigeration work, with all 'hard' tubing joints silver soldered or brazed.

Type L – 'medium' wall thickness, made in 'hard' and 'soft' grades. Suitable for plumbing and heating work.

Type M – 'light' wall thickness, and available only in 'hard' grade. Suitable for drainage lines.

N.B. – the above types are supplied in straight lengths, the ends of which should be capped.

Refrigeration quality – supplied in dehydrated and sealed coils in OD sizes up to and including $1\frac{3}{8}''$. Is soft-annealed to permit flaring and assist bending.

Wall sizes vary in accordance with OD dimension:

a) Refrigeration quality coiled tubing:

Wall thickness	OD
0·030''	$\frac{1}{4}''$
0·032''	$\frac{1}{2}''$
0·035''	$\frac{3}{4}''$
0·045''	$\frac{7}{8}''$

b) Type K:

Wall thickness	OD
0·035''	$\frac{3}{8}''$
0·049''	$\frac{3}{4}''$
0·065''	$1\frac{1}{8}''$
0·072''	$1\frac{5}{8}''$
0·083''	$2\frac{1}{8}''$
0·095''	$2\frac{5}{8}''$
0·109''	$3\frac{1}{8}''$
0·134''	$4\frac{1}{8}''$

Fig. 14.2 Copper tubing wall thickness

Steel tubing dimensions

Steel tube sizes have traditionally been quoted in terms of *internal* diameter. A number of wall thicknesses have been and are available for differing duties, and have a direct bearing on the pressures which pipes can withstand.

The word 'pressures' must itself be qualified, since limits change with the temperatures of liquids or gases flowing through the pipes. For the majority of chilled and condenser water lines, 'medium weight' tubing and fittings are adequate, whilst R717 lines must be in 'heavy' tubing. However, when water circuits are exposed to high static heads or system resistances/pumping heads, selections must be made on the basis of proper calculations, not rules of thumb. This consideration applies also to jointing methods – screwed joints, especially in large diameter lines, are not suitable for high pressure work; and weld flanges can also present problems when used with high pressure fluorocarbon refrigerants.

ISO pipe size designations are:

ISO mm	Actual OD mm	Nominal bore (inches)
15	21	$\frac{1}{2}$
20	27	$\frac{3}{4}$
25	34	1
32	43	$1\frac{1}{4}$
40	48	$1\frac{1}{2}$
50	60	2
65	76	$2\frac{1}{2}$
80	89	3
100	114	4

Fig. 14.3 Steel pipe dimensions

ISO dimensions are quoted in suppliers' literature and on system drawings. It is essential to remember this when ordering pipe insulation, which has to fit the OD, not the nominal (internal) size of the tubing!

Sheet metal gauges

The student may also need to be able to relate the thickness, or gauge, of sheet metal in terms of both millimetres and Birmingham Gauges:

mm	BG
0·50	26
0·63	24
0·79	22
1·00	20
1·26	18
1·59	16
2·00	14
2·51	12

Fig. 14.4 *Sheet metal thicknesses*

SWG and American gauge thicknesses are slightly different from those quoted above, and from each other. We have quoted BG thicknesses in this instance because they tend to fall between dimensions of other systems, and the differences are not in any case major ones.

Glossary

Absorption. A process incurring the removal of one or more constituents of a mixture of liquids or gases by another material which undergoes physical or chemical change in the process.

Adsorption. A process incurring the removal of one or more constituents of a mixture of liquids and gases by surface adherence to another material which does not undergo any physical or chemical change.

Air changes. The introduction of fresh ambient air to an air-conditioned or refrigerated space; expressed in terms of complete air changes in the space per unit period of time.

Air-conditioning (close control). The process of accurately controlling air cleanliness, temperature and relative humidity to meet specific equipment or process requirements.

Air-conditioning (comfort cooling). The process of controlling air cleanliness and temperature without closely regulating its relative humidity, to increase the comfort of occupants of the treated space.

Ambient air. The air surrounding a space or building structure being air-conditioned or refrigerated.

Anemometer. An instrument to measure the velocity of a moving fluid.

Azeotrope. A mixture of two refrigerants which behave as one substance when they evaporate or condense.

Barometer. An instrument to measure atmospheric pressure.

Bimetallic element. An element made from two metals with differing coefficients of thermal expansion, used in a thermal control device.

Boiling point. The temperature at which the vapour pressure of a liquid is equal to the absolute pressure of the atmosphere in contact with the surface of the liquid.

Brazing. Joining two metallic surfaces using a hard solder or similar material which is drawn by capillary action into a fine gap between the surfaces to be bonded.

British thermal unit (Btu). The amount of heat required to raise the temperature of 1 lb of water through 1 °F.

Bulkhead. A dividing wall.

Calorie. The amount of heat required to raise the temperature of 1 g of water through 1 °C.

Capillary tube. A tube of small internal bore, used either to meter the flow of liquid refrigerant into a cooling coil or to transmit pressure from the sensing phial of a control to its operating element.

Carbonisation. The formation of deposits of carbon, normally produced by the decomposition of oil as the result of overheating.

Cascade system. One having two separate refrigerant circuits, arranged so that the evaporator of one removes heat from the condenser of the other.

Change of state. The process of conversion of matter from solid to liquid or liquid to gas, or vice versa.

Coefficient of performance – COP (refrigeration compressor). The ratio of refrigerating effect to the energy input to the shaft.

(heat pump). The ratio of the heating effect delivered to the energy input to the equipment.

Commutator. A split ring which is used to reverse the contacts of a generator every half turn, thus allowing a d.c. rather than an a.c. current to be generated.

Compression, compound. Compression of refrigerant vapour in two or more cylinders, in stages.

Compression ratio (refrigerant). The ratio of absolute pressures before and after compression.

Condensation. The conversion of a vapour into a liquid by the removal of latent heat. Occurs in the condenser of a refrigeration system, or on the surface of a cooling coil at a temperature below the Dew Point of the air.

Condenser. Air-, water- or evaporative-cooled heat exchanger in which heat is extracted from high pressure refrigerant vapour which then condenses and

217

assumes its liquid state.

Conduction (thermal). The process of transfer of thermal energy between sub-molecular particles of matter in contact with each other.

Conductor (electrical). A substance which allows an electric current to flow through it.

Convection. A pattern of movement in liquids or gases, resulting from density variations caused by temperature changes.

Cryogenics. The science of obtaining extremely low temperatures and investigating their effects on matter.

Current. A flow of electrons.

Defrosting. The removal of accumulated ice from the surfaces of cooling coils which operate below freezing point.

Dehumidification. The removal of water vapour from air; normally performed by passing air through a finned heat exchanger at a temperature lower than Dew Point, but can also be achieved chemically.

Density. The mass per unit volume of a substance.

Desiccant. A chemical which attracts water or water vapour and removes it from its surroundings. Desiccants widely used in the refrigeration industry include activated alumina, silica gel and molecular sieve.

Dew Point temperature. The temperature at which the condensation of water vapour contained in air will commence as the temperature level is reduced. It corresponds to the saturation point of any given mixture of air and water vapour.

Direct expansion. The process of heat removal by the evaporation of a refrigerant in a cooling coil (evaporator).

Drop out current. The voltage at which a solenoid coil allows the armature to drop and break the circuit.

Dunnage. Strips of timber used in cold stores to separate walls from packages, or packages from each other, to provide space for the circulation of air.

Earthing. The practice of connecting all non-current-carrying metal to earth. It prevents a dangerous build-up of voltage if there is an electrical fault.

Electromotive force (e.m.f.). Applies to electrical cells – the total work done per coulomb of electrical charge conveyed in a circuit in which the cell is connected. It is the force that makes electrons move in a circuit.

Enthalpy. The heat content of a substance – most particularly of air and its associated water vapour.

Evaporation. The change of state from liquid to vapour during which heat is absorbed from surrounding space or matter.

Evaporator. An instrument which absorbs heat from its surroundings by the evaporation of liquid refrigerant.

Expansion valve. A device designed to regulate the flow of refrigerant.

Flammability. The ability of a substance to burn.

Flash point. The temperature at which a combustible substance, such as oil, has been vapourised to an extent which will support combustion.

Freezing point. The temperature at which a liquid is converted to its solid state, upon the removal of its latent heat of fusion.

Frost back. Flooding of liquid refrigerant from an evaporator into the suction line, which usually then frosts over.

Fumes. Airborne particles of solid matter, usually less than 1 micron in diameter, normally resulting from condensation or chemical reactions.

Fusible plug. A device to relieve pressure, by melting at a predetermined temperature.

Gasket. A door sealer, usually magnetic with a rubber or plastic casing.

Head, static. The static pressure of a fluid, expressed in terms of the height of a column of manometric fluid which it would support.

   total (or **dynamic**). The total of static head and velocity head pressures.

   velocity. The height of a column of manometric fluid equivalent to the velocity pressure of a moving fluid.

Heat conductor. A substance which will readily permit the passage of thermal energy. The opposite of a thermal insulator.

Heat exchanger. A device designed to transfer heat between two substances which are physically separated from each other.

Heat, latent. The amount of heat required to change the state of a substance (e.g. from a liquid to a gas) without changing its temperature. It is applicable to the processes of freezing, boiling, condensation and melting.

Heat pump (cooling and heating). A refrigeration system which can be used either to cool or to heat a given space, normally by exchanging the functions of the evaporator and the condenser.

Heat pump (heating). A refrigeration system designed specifically to remove heat from matter, to increase its intensity and to utilise all the heat rejected by the system for a heating function.

Heat, sensible. A measure of the intensity of heat content which can be measured on a normal thermometer.

Hermetic compressor. A unit which contains the motor and compressor sealed into the same casing.

High side. That part of a refrigeration system which operates at about the same pressure as its condenser.

Horsepower. A unit of power in the British system, equivalent to 745·7 W and normally applied to a motor or other form of prime mover.

Hot gas line. A pipe used to convey discharge gas from a compressor to the entrance of the condenser.

Humidity. The presence of water vapour within a space or air mass.

Humidity, relative. The ratio of the mass of water vapour present in an air mass to the mass of water vapour which would saturate the same volume of air at the same sensible temperature. It can also be expressed as the ratio of the partial pressure of water vapour in the air to the saturation pressure of water vapour at the same temperature.

Hygrometer. An instrument which indicates, by the extent to which it sinks within a container of liquid, the specific gravity of the liquid.

Hygroscopic. Readily absorbs moisture.

Infiltration. Air flowing into a space through gaps around doors, windows, etc.

Insulator, acoustic. A substance used to reduce the amount of noise and/or vibration through a system or structure.

Insulator, electrical. A substance which does not allow an electric current to flow through it.

Insulator, thermal. A substance which offers considerable resistance to the flow of heat.

Liquid line. A pipe used to convey liquid refrigerant from the receiver or condenser of a system to its refrigerant control device.

Liquid receiver. A vessel inserted in a system to contain the entire charge of refrigerant if the system is pumped down.

Low side. That part of a refrigeration system which operates at about the same pressure as its evaporator.

Manifold. Part of a main pipeline at which several branch lines are connected in close proximity to each other. Alternatively, a single main pipeline incorporating several separate flow paths.

Manometer. An instrument used to measure pressures – generally a U tube partially filled with a fluid which indicates, when displaced, the amount of pressure exerted on it.

Mass. The quantity of matter in a body, measured in grams and kilograms in the SI system and in pounds in the British system.

Micron. One thousandth of a millimetre.

Ohmmeter. An instrument to measure the electrical resistance of circuits or parts of circuits.

Organic compound. Conventionally a chemical compound produced by the processes of living organisms, but now applied to nearly all compounds containing carbon.

Performance factor. The ratio of the useful capacity of a system to the energy input required to obtain it.

Pilot light. A small flame kept alight to light another.

Potential difference (p.d.). The difference in energy content of electrons at two points. Measured in volts.

Power. The rate at which work is performed. Measured in watts.

Pressure (absolute). Pressure expressed so that a perfect vacuum has a value of zero. The sum of atmospheric and gauge pressures.

(gauge). Pressure above that of the atmosphere. At atmospheric pressure the gauge reading should be zero.

(saturation). The pressure at which the solid and liquid phases of a pure substance are in equilibrium.

(static). The pressure on a surface at rest in relation to the fluid flowing over it.

Pressure drop. A reduction in the static pressure of fluid flowing through a pipe or duct, due to friction etc.

Psychrometer. Instrument for measuring relative humidities by means of dry and wet bulb temperatures.

Pull in current. Voltage at which solenoid coil attracts an armature to make the circuit.

Pyrometer. An instrument for measuring very high temperatures.

Refrigerant control. See **expansion valve** and **capillary tube**.

Resistance (electrical). The opposition a circuit offers to the flow of electricity. Measured in ohms.

(thermal). Resistance to the flow of heat.

Respiration. The production of carbon dioxide and heat by stored fruit and vegetables as the result of ripening.

R.H. See **humidity**.

Saturation. A condition in which vapour cannot support additional moisture content unless its temperature is increased. The vapour and liquid are in equilibrium.

Scale. Deposits left behind by water which may coat pipes and other surfaces and reduce efficiency.

Sensible heat ratio. The ratio of sensible cooling to total cooling load or effect.

Shims. Thin strips of material used to take up the spaces between parts.

Sludge. A by-product of the decomposition of oil, resulting from chemical reactions or the presence of moisture or other impurities, especially at moderate to high temperatures.

Specific heat capacity. The ratio of the heat required to raise the temperature of a mass of a substance through a measured temperature range to that required to raise the temperature of the same mass of water by the same amount.

Stator. The stationery part of a generator or motor.

Steam, saturated. Steam at its saturation temperature and carrying water particles in suspension. (Steam at saturation temperature but not containing water particles is described as **dry saturated**.)

Subcooling. Cooling liquid refrigerant below the temperature equivalent to the system condensing pressure; or cooling a liquid such as water to a temperature below its normal freezing point, without changing its state.

Sublimation. A change of state direct from solid to gas, without the formation of a liquid.

Suction line. A pipe carrying suction gas from the evaporator to the compressor inlet valve or port.

Superheated. (Of refrigerant or steam) at a temperature higher than the saturation temperature that is equivalent to a given pressure.

Swarf. Metal chips and turnings removed by cutting tools during machining operations.

Temperature (absolute zero). Zero on the kelvin scale, equalling $-273 \cdot 16\ °C$.

(dry bulb). The temperature as indicated by an accurate thermometer, of which the bulb is affected only by the medium being measured.

(wet bulb). The temperature indicated by an accurate thermometer, the bulb of which is contained within a saturated wick from which water evaporates into the air.

Thermal transmittance (U factor). The rate of heat flow per unit area from the warm to the cooler side of a structure which offers resistance to the flow of thermal energy.

Ton of refrigeration. A conventional British unit of refrigeration capacity, equalling the refrigerating effect of one short ton of ice melting over a period of 24 hours. 12 000 Btu/h equals 3 516 W. (A 'short ton' is 2 000 lb; a 'long ton' is 2 240 lb.)

Vacuum. A pressure less than that of atmospheric pressure.

(perfect). An absolute pressure of zero.

Vapour. A gas which is approaching the liquid phase of the substance, and does not follow ideal gas laws.

Vapour barrier (or seal). A seal or membrane which is impervious to moisture; applied to the warmest or most humid face of a structure.

Vapour pressure. The pressure exerted by a vapour, either by itself or in a mixture of gases.

(saturated). The vapour pressure of a vapour in contact with its liquid form. It increases with a rise in temperature.

Viscosity. Resistance to any change of shape or arrangement which causes friction when adjacent layers of fluid move in relation to each other.

Volume, specific. Volume per unit mass – the reciprocal of density.

Water gauge. A scale used for measuring pressure.

Water vapour. Normally refers to air containing evaporated water, or steam.

Wax. A solid substance which may be deposited if a mixture of oil and refrigerant is cooled.

Wet bulb depression. The difference between dry bulb and wet bulb temperatures.

White room. A space which is free from measurable quantities of dust or bacteria.

Wiredrawing. A restriction in the area of

a pipe available for the flow of a fluid; it throttles flow and reduces fluid pressure.

Abbreviations of organisations

AHAM	Association of Home Appliance Manufacturers
ANSI	American National Standards Institute
ARI	Air-conditioning and Refrigeration Institute
ASHRAE	American Society of Heating, Refrigerating and Air-conditioning Engineers
ASTM	American Society for Testing and Materials
CIBS	Chartered Institute of Building Service Engineers
HCVA	Heating and Ventilating Contractors' Association
ISO	International Standards Organisation
SAE	Society of Automotive Engineers
SMACNA	Sheet Metal and Air-conditioning Contractors' National Association
UL	Underwriters' Laboratories

Examination papers

CITY AND GUILDS OF LONDON INSTITUTE

PAPER NUMBER 827-2-05	EXAMINATION REFRIGERATION II (INDUSTRIAL REFRIGERATION)	YOU WILL NEED
SERIES MAY–JUNE 1979	PAPER SECOND WRITTEN TIME ALLOWED 2 HOURS	answer book drawing instruments

The maximum mark for each question is shown.
Answer TWELVE of the following fourteen questions, ALL in Section A and TWO from Section B.
Allow about 1 HOUR for each section.

SECTION A – Answer ALL the questions.
1. On a service call, an hermetic compressor is found to start, run for a short time, then cut out on high pressure control. There are no electrical defects. List FOUR possible defects in the refrigeration system.
(5 marks)
2. Why is the diameter of a liquid line drier always larger than that of the tubing in which it is installed? (5 marks)
3. The condensing pressures recorded in the operating log of a water-cooled condensing unit are very different from those recorded when the system was first installed. List THREE factors which could cause such variations. (5 marks)
4. Name the refrigeration circuit component shown in Fig. 1 and describe briefly how it operates. (5 marks)
5. List FIVE possible causes for faulty operation of a belt-driven condenser fan, if the fan motor is not defective. (5 marks)
6. An evaporator is to operate at a temperature of 40 °F, using R22 as the refrigerant. There are pressure drops of 9 lbf/in² gauge through the refrigerant distributor and 3 lbf/in² gauge through the evaporator itself. At what pressure should refrigerant leave the expansion valve? (At 40 °F the vapour pressure of R22 refrigerant = 69 lbf/in² gauge)
(5 marks)
7. List FIVE contaminants which must be kept out of the refrigeration system during its installation. (5 marks)

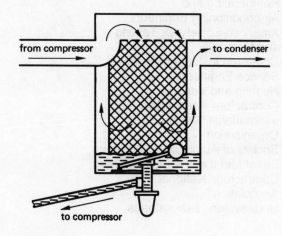

Fig. 1

8. At which TWO points in a refrigeration plant are measurements taken to determine the superheat of the refrigerant? (5 marks)
9. Which TWO factors determine the size and layout of refrigerant lines in any installation? (5 marks)
10. What are the TWO main reasons for vapour-sealing a cold room?
(5 marks)

EXAMINATION PAPERS

SECTION B – Answer TWO questions only.

11 (a) List the sequence of events when the demands on a refrigeration system having a pumpdown control require the compressor
   (i) to start; .... (7)
   (ii) to stop. .... (7)
   (b) Why could the compressor be started by the pressure control, when cooling is not called for by the thermostat? .... (6)
   (c) What additional control can be fitted to prevent this and how does it operate? .... (5)
   (25 marks)

12 (a) Make a sketch showing an oil pressure safety control incorporated in a refrigeration compressor. .... (10)
   (b) Label the main components of the control on your sketch. .... (5)
   (c) Describe how the control operates. .... (10)
   (25 marks)

13 (a) Make a layout sketch of a block ice-making plant of the brine-tank type using a water-cooled condenser. .... (10)
   (b) Label the main components on your sketch. .... (5)
   (c) List THREE types of condenser which could be used. .... (5)
   (d) State the routine precautions to avoid corrosion of the brine tank. .... (5)
   (25 marks)

14 List the steps to be taken to replace the stator of a large serviceable hermetic compressor after a motor burnout. (Refer to the hermetic unit only). .... (25 marks)

223

TROPICAL REFRIGERATION AND AIR-CONDITIONING

CITY AND GUILDS OF LONDON INSTITUTE

PAPER NUMBER 827–2–07   EXAMINATION **REFRIGERATION II (AIR-CONDITIONING)**
SERIES **DECEMBER 1979**   PAPER **AIR–CONDITIONING**   TIME ALLOWED **2 HOURS**
YOU WILL NEED **one answer book   drawing instruments**

Answer TWELVE of the following fourteen questions, ALL in Section A and TWO from Section B.
Each question carries 5 marks in Section A and 25 marks in Section B.
Allow about 1 hour for each Section.

SECTION A – Answer ALL the questions. Each question in this Section carries 5 marks.

1. List THREE types of safety control used to guard against freeze-up when packaged water chillers are used in air-conditioning installations.
2. (a) Name THREE items of equipment which are contained in an air-conditioner used to maintain precise conditions in a computer suite but not contained in a simple packaged air-conditioner used for comfort cooling.
   (b) What condition does this equipment control?
3. State briefly why the thermostatic expansion valves used on direct-expansion air-conditioning installations are normally equipped for external pressure equalisation.
4. Describe the normal method by which the motor of a serviceable hermetically sealed centrifugal compressor is protected from refrigeration overload.
5. List FIVE reference lines or curves normally appearing on a psychrometric chart.
6. (a) Name the type of control normally used to start and stop a steam pan humidifier.
   (b) State TWO places where such a humidifier may be installed in an air-conditioning system.
7. (a) Name the instrument normally used to measure pressures in low-pressure air-distribution ductwork.
   (b) State why its scale is normally inclined from the vertical.
8. List the THREE controls incorporated in a direct-expansion air-conditioning plant to provide a refrigerant pump-down system.
9. (a) Describe briefly the position of the refrigerant cylinder when used to add a small amount of refrigerant to a system.
   (b) Give TWO reasons for this positioning of the cylinder.
10. Describe briefly:
    (a) when an acid testing kit should be used on a serviceable hermetically sealed compressor assembly;
    (b) how it should be used.

SECTION B – Answer TWO questions only. Each question in this Section carries 25 marks.

11. Make a sketch showing the general layout and name the main components of EACH of the following units when incorporated in a packaged water chiller powered by a centrifugal compressor:
    (a) a purge unit;   (b) a purge drum.
12. Describe the most likely reason for EACH of the following fault symptoms in a centrifugal fan, and its drive, as incorporated in an air-handling unit. State what action should be taken to correct the fault in EACH case:
    (a) fan vibration;   (b) rumbling noise from the fan section;
    (c) squealing noise from the drive;   (d) fan pulsation (or 'beating').
13. List in their correct sequence the ELEVEN main steps to be taken when it is found necessary to replace a large serviceable hermetically-sealed reciprocating compressor following a burn-out of the compressor motor, assuming that the system refrigerant is to be used again. Include where necessary the checks required during repair and any testing afterwards.
14. State in detail the effects on a central plant-type air-conditioning installation of EACH of the faults as listed below:
    (a) FOUR effects of the tubes in the water-cooled condenser being fouled with sludge and scale;
    (b) FOUR effects of the air filters in the chilled water system fan/coil units being clogged with dust;
    (c) TWO effects of poor suction line design preventing the proper return of oil to the compressor;
    (d) FIVE effects of pipe scale being carried through chilled water piping not fitted with strainers;
    (e) TWO effects of the freeze protection thermostat on the water chiller being set 3 °C (or 5 °F) too low.

CITY AND GUILDS OF LONDON INSTITUTE

PAPER NUMBER 827-1-02	EXAMINATION **REFRIGERATION I**	YOU WILL NEED
SERIES **MAY–JUNE 1979**	PAPER **TRADE STUDIES (2)** TIME ALLOWED **2 HOURS**	**one answer book** **drawing instruments**

Answer TWELVE of the following fourteen questions, ALL in Section A and TWO from Section B.
Each question in Section A carries 5 marks; each question in Section B carries 25 marks.
Allow about 1 hour for each Section.

SECTION A – Answer ALL the questions. Each question in this Section carries 5 marks.

1. State why the interior of refrigerant systems must NOT be flushed out with oxygen during service operations.
2. (a) State why domestic refrigeration systems are normally topped up with gaseous refrigerant.
   (b) State how the refrigerant cylinder is positioned to prevent liquid refrigerant entering the system.
3. A safety thermostat is wired in circuit with an electric heating coil forming part of a domestic air-conditioner. State TWO faults affecting different components which would cause the thermostat to operate.
4. List THREE symptoms which could result from a defective door gasket on a domestic refrigerator.
5. For EACH of the following types of evaporator, name ONE refrigeration system in which the evaporator would be used:
   (a) embossed plate;   (b) coil and tank;   (c) cross finned.
6. A refrigeration system is undercharged with refrigerant. Describe the symptoms which would be seen through the liquid line sight glass.
7. State the effects of air starvation on a domestic refrigeration system incorporating an air-cooled condenser with reference to:
   (a) head pressure;   (b) compressor motor amperage;
   (c) compressor operating temperature.
8. (a) State where a vapour barrier is applied when insulating a low temperature coldroom.
   (b) State the reason for having a vapour barrier.
   (c) Describe ONE safety device which should be present in the door hardware for a large coldroom.
9. List THREE possible causes of internal damage to ventilated frame electric fan motors which may be prevented by routine maintenance.
10. State briefly why the room air processed by a domestic dehumidifier is first cooled and then heated.

SECTION B – Answer TWO questions only. Each question in this Section carries 25 marks.

11. (a) With reference to oils used in refrigerators, define briefly:
    (i) viscosity;   (ii) miscibility.
    (b) State why and how the miscibility of oil can affect the design of refrigerant pipelines in R22 installations.
    (c) Describe how a hermetically-sealed refrigerant compressor is lubricated, making a line sketch of its internal construction to illustrate your answer.
12. (a) Make a line sketch of a domestic-type room air-conditioner and label the main components.
    (b) (i) Explain how the room condensate accumulating in the unit baseplate is disposed of by the condenser fan.
        (ii) State why this method of disposal is used.
13. (a) List in their correct sequence the main steps to be taken when removing and replacing the hermetic compressor of a domestic air-conditioner and re-charging the system following the burn-out of the compressor motor.
    (b) Explain why an acid-testing kit is NOT used after this type of repair.
14. (a) Make a labelled sketch to show a charging board connected to the charging valve of a refrigerator and to a cylinder of R12. Include liquid refrigerant scale, vacuum pump, refrigerant lines, refrigerant valves and gauges.
    (b) With reference to your sketch, explain briefly how the charging board is used to evacuate and charge the refrigerator.

Index

note: page numbers in italics refer to illustrations

absolute pressure, 10–11
absorption refrigerators, 1, 38, 58–9, 69–70, 71, 116, 171
absorption water chillers, 116, 118–21
accessible hermetic compressors, 26, 73, *74*, 76, 111, 121, 122, 144, 191–4
across line starters, *see under* starters
adsorption charges, 91
air, properties of, 9, 14–16
air-cooled condensers, 13–14, 52, 73, 79–80, 86, 87, 100, 105, 111, 118, 122
air distribution, 133–4, 155–8
air grilles and diffusers, 155–7
airhandling units, 127–30, 172–4, 177–9
air mixing, 128
air pressure, 9, 10, 11, 14, 98
air registers and diffusers, 134
air velocities, 157
alternating current (a.c.), 22, 23
alternating current (a.c.) motors, 24–5, *31*, 32–3;
  see also single-phase a.c. motors; three-phase a.c. motors
ammeter, 19
  clamp-on, 200
ammonia, 1, 2, 27, 41, 58–9, 119;
  see also R717
amperes (A), 19
anti-vibration mountings and foundations, 158–60
atmospheric pressure, 9–10, 11, 14

atomic structure, 17
auto-transformer starters, *see under* starters
axial-type fans, 133–4
azeotrope, *see* R500; R502

bellows-type 'motor', 80
blast freezers, 106, 107
bleeding, treatment for, 211–12
block ice plants, 35, 104–6
boiling point, 4, 11, 12
Bourdon-type pressure gauge, 10
brines, 43, 45, 93, 104–6
British thermal units (Btu), 4
bubble tests, 41
burns, treatment for, 35, 211

cabinets
  for air-conditioners, 59–60
  for airhandling units, 130
  for refrigerators and freezers, 46–7, 59, 71, 72–3, 98–101
  see also display cases, refrigerated
calories (cal), 4
capacitance, 23–4
capacitor analyser, 200, 203–4
capacitors, 23–4
capacitor start, capacitor run (C.S.R.) motors, *see under* single-phase a.c. motors

capacitor start, induction run (C.S.I.R.) motors, *see under* single-phase a.c. motors
capacitor tests, 202–4, 208
capacity control devices, 75–6, 108, 121, 122
capillary tubes, 28, 50, 53, 61, 77, 80, 89, 125, 189
Carré, Ferdinand, 1
carrene 500, *see* R500
Carrier, Willis, 1
cascade systems, 109
celsius (centigrade) scale, 4–5
central plant, 116
centrifugal chillers, 116–19
centrifugal compressors, 35, 37, 116, 117–18, 121, 122, 126
centrifugal fans, 129–30, 134, 135, 136
charging procedures, 144–7, 195–6
chilled water, 43–4, 116
chilled-water circuits, 151–2
chlorodifluoromethane, *see* R22
chloropentafluoroethane, *see* R502
circuit breakers (C.B.), 33, 204
circuit tester, 199
cleanliness, 138–9
close-control air-conditioning, 112–15, *116*
codes and standards, 62–3, 67
coil temperature, 112
cold rooms and cold storage, 2, 35, 98, 101–4

226

INDEX

colour coding, 33, 36, 37–8
comfort-cooling air-conditioning, packaged, 110–12, 114, 115, *116*;
  see also room air-conditioners
commercial and industrial air-conditioners, 35, 43, 71, 110–37
commercial and industrial refrigerators, 28, 35, 37, 71–109, 169–71
commutators, 22, 24–5
compression ratios, 43, 48–9, 73, 108
compressor efficiency, 49
compressor overloads, 57, 62
compressors, 2, 26, 42, 59, 65, 73–9, 85–6, 139;
  see also centrifugal compressors; hermetic compressors; motor tests; open compressors; rotary compressors; screw compressors; welded hermetic compressors
compressor safety controls, 76–7
condensate disposal, 65–6, 67, 69, 71, 99, 129, 137
condenser capacities, 85–6
condenser design trends, 87
condensers, 44, 51–2, 59, 61, 64, 65, 66, 69, 73, 79–87, 122, 139, 180–3
  forced-draught, 52, 62, 71, 79, 82, 125
  natural-draught, 52
  skin-wound, 52
  see also air-cooled condensers; water-cooled condensers
condenser water circuits, 152–3
conduction, 7–8
conductors, 17, 19–21, 33
constant pressure expansion (C.P.) valves,
  see under valves
contaminants, 89, 139, 189–90
controls
  adjustment of, 188–91
  humidistat, 66–7

refrigerant, 13, 14, 89–92, 144–5;
  see also capillary tubes; valves, expansion,
  room unit, 62
  see also safety precautions/controls; system controls in refrigeration circuits
convection, 8
conversion factors, *5*, *6*, 213–16
cooling coils, 66, 71, 90, 91, 101, 105, 128–9, 189
cooling loads, refrigeration, 103–4
copper tubing, 141–2, 215
coulombs, 19
Cullen, William, 1
current electricity, 18–20
cylinders, 36, 47–8, 50, 75, 76, 145

dairy cases, 100
Dalton's law of partial pressure, 14
defrosting, 53–4, 56, 64, 67, 69, 71, 92–3, 99, 100
dehumidifiers, 66–7, 69
dew point temperature, 14, 72, 73
dichlorodifluoromethane, *see* R12
dichlorotetrafluoroethane, *see* R114
difluoroethane, *see* R500
direct current (d.c.), 22
direct current motors, 24, 28–30, *31*, 32
  compound wound, 28, *29*, 30
  series wound, 28, *29*
  shunt wound, 28, *29*, 30
direct drive couplings, 162–3
direct-on-line starters, *see under* starters
discharge pressure, 48, 111
display cases, refrigerated, 98–101
domestic air-conditioners, 70, 82;
  see also room air-conditioners; split system air-conditioners
domestic freezers, 46–59
domestic refrigerators, 28, 33, 35, 37, 38,

46–59, 67, 69–70, 144, 201;
  see also absorption refrigerators
drainage lines, 153
drinking water coolers, 125–6
drive kit installation, 160–3
dry bulb temperature (DB), 14, 63, 82, 98, 111, 112, 197
ductwork installation, 138–63, 164–5

earthing, 33
effective temperatures (ETs), 111
electrical energy, 24
electrical faults, 168
electrical service procedures, 199–209
electric heating elements, 63
electricity, 17–34
electric shock, 210–11
electro-chemical generation, 18
electro-magnetic generation, 18–19;
  see also generators; motors
electromotive force (e.m.f.), 19, 20, 21
electronic air cleaners, 136
electronic leak detectors, 41, *42*
electronic refrigeration circuit, 33–4
electron theory, 17–18
energy, 7;
  see also electrical energy; mechanical energy; solar energy
enthalpy, 14, 213
evacuation procedure, 43, 142–3
evaporative air coolers, 135–6
evaporators, 13, 42, 43–4, 53–4, 59, 64, 66, 76, 89, 90, 91, 92–6, 98, 112, 189
  bare pipe air cooler, 94
  Baudelot cooler, 93–4
  forced-draught, 71, 92–3, 98, 99–100, 105
  shell and tube, 93, 94, 118
  submerged, 93
  see also controls, refrigerant,
expansion valves, *see under* valves

227

fahrenheit scale, 5
fan/coil units, 130, *131*
fans, 26, 28, 30, 62, 79, 80, 82, 107, 110, 133–4, 208;
  see also centrifugal fans
Faraday, Michael, 20
farads (F), 23
faults, 168–87
filtration, 128
fire point, 43
first aid, 210–12
fish display cases, fresh, 100–1
flash point, 43
float valves, 79, 89, 91–2, 93, 105, 119, 124
  high-side, 92
  low-side, 92
flux lines, 20–1
food freezing, 35, 106–7
food storage, 96–8
four-wire, three-phase installations — a.c., 32–3
freezing point, 4
Freon piping practices, 147–9
frost heave, 102
fruit and vegetable display, 100
fuse puller, 200
fusible plugs, 83–4

gas charges, 91, 195
gas lines, pitching horizontal, 148
gauge pressure, see pressure gauges
generators, 19, 21–3, 24
glycol, 43, 44–5
Gorrie, John, 1

halide test lamps, 41
Harrison, James, 1
head pressure controllers, 190, *191*, 196

heat, 4, 213;
  see also enthalpy; latent heat; sensible heat; specific heat capacity
heat exchangers, see condensers
heat generation, 18
heating coils, 129
heating unit systems, 63–5;
  see also pumps, heat,
heat transfer, 7–9, 105, 215
hermetic compressors, 26, 27, 28, 42, 61, 139;
  see also accessible hermetic compressors; reciprocating hermetic compressors; rotary compressors; welded hermetic compressors
hermetic unit analyser, 201, 205–6
Hertz (Hz), 22
high potential tester, 200–1, 205
high velocity induction units, 130–2
history of air-conditioning and refrigeration, 1–3
horsepower, 20, 214
hot gas bypass valves, see under valves
humidifiers, 112, 113–14
  electrode, 114
  steam pan, 113–14
  water atomising, 114, *115*
humidistats, 66–7, 112–13, 114
hydrometer, 106

icemakers, 106;
  see also block ice plants
induction start, induction run (I.S.R.) motors, see under single-phase a.c. motors
installation, 67–9, 164–6
insulation, 16, 30–1, 46, 50, 71, 72–3, 101, 123, 133, 149, 152, 155

joules (J), 4

Kelvin, Lord, see Thompson, William,
Kelvin scale, 5
Klixon, 57

latent heat, 7, *8*, 12, 14, 36, 59, 97, 103, 213
Law of Conservation of Energy, 1, 7
leak detection, 41–2, 143–4, 150–1
Legionnaires' disease, 136–7
light generation, 18
liquid charges, 91
liquid lines and fittings, 52–3, 65, 79–80, 87–9
liquid-line strainer/driers, 52, 87–8, 89, 189, 190
liquid receivers, 83, 87
litmus paper, 41
log sheets, 165–6
low temperature displays, 99–100
low temperature refrigeration systems, 107–9
lubrication, see oil

magnetism and electricity, 20–1
maintenance, 69–70, 166–7
manometer ('U' tube), 10–11
meat display cases, 98–9
mechanical energy, 24
'meggers', 200, 202, 204–5, 207, 208
mercury barometer, 9
Midgely, Thomas, 2
miscibility, 43
moisture, 14, 42–3, 46, 72, 87–8, 98, 139, 142–3
moisture indicators, 42, 88
motor faults, *31*, 187
motor frames, 30
  drip-proof, 30
  flame-proof, 30
  screen protected, 30

INDEX

totally enclosed, 30
totally enclosed, fan-cooled, 30
motor insulation, 30–1
motors, 24–31, 50, 73, 74, 76–7, 79, 85, 110, 194, 207;
  see also direct current motors; fans; single-phase a.c. motors; three-phase a.c. motors
motor tests, 20, 204–7

ohmmeter, 20, 200, 207, 208
ohms (Ω), 19
Ohm's law, 19–20
oil, 50, 73, 74, 76, 108, 139, 150, 189, 193, 194
  entrained, 43, 77
  refrigeration, 42–3
  see also pumps, oil; sight glasses
oil charges, 146–7
oil pressure control, 73, 77, *78*, 118
oil return, 147
oil separators, 77–9, 119, 150
oil traps, 147–8
open compressors, 27, 28, 30, 38, 42, 73–4, 76, 85, 107–8, 144, 191–4;
  see also rotary compressors

part line starters, see under starters
pascal (Pa), 9
Peltier effect, 33–4
Perkins, Jacob, 1
permanent split capacitor (P.S.C.) motors, see under single-phase a.c. motors
'phase, in', 24
'phase, out of', 24
pipelines, 42, 138–63, 164
plate freezers, 106, 107, *108*
potential difference (p.d.), 19–20, 23, 42
pour point, 43
power, active, 24

power, electrical, 20, 24, 214
power distribution, 31–3
power factor, 24
pressure, 9–12, 53, 90–1, 214
  high condensing, 195–6
  in welded hermetic systems, 196–7
  see also air pressure; atmospheric pressure; discharge pressure; suction pressure; *and under* oil
pressure controls, 73, 76–7, 190, *191,* 196
pressure gauges, 9–11, 12, 38, 196
pressure relief devices, 80, 83–5, 87;
  see also under valves
pressure/temperature relationships, 11–12, 38, *39–40*, 90
psychometric chart, 14–16, 66, 98
psychrometer, sling, 14
pumps, 26, 82, 88, 105, 118, 150, 189
  booster, 43, 109
  condensate, 67
  heat, 3, 65, 111
  oil, 73, 77, 118
  refrigerant, 109
  vacuum, 142–3, 146
  water, 82, 126–7, 153–4
purge units, 118–19, 120

radiation, 9
Rankine scale, 5–6
reach-in refrigerators, 71–2, 89
reciprocating hermetic compressors, 1, 37, 38, 47–8, 49–50, 73, 74, 76, 85, 107–8, 121;
  see also hermetic compressors
reduced voltage starters, see under starters
refrigerant charging, 144, 195–6
refrigerant controls, see under controls
refrigerant pipelines, 139–42
refrigerants, 2, 12, 13, 35–45, 50, 188

R11, 35, 37, 38, 77, 117, 193, 195
R12, 2, 12, 35, 37, 41, 42, 50, 71, 73, 74, 77, 95, 104, 106, 108, 117, 125
R22, 35, 37, 38, 42, 50, 61, 73, 74, 77, 95, 108, 147
R114, 37, 51
R115, 38
R152a, 37
R500, 37–8
R502, 38, 42, 73, 74, 77, 95, 100, 104
R717, 35, 38, 42, 43, 74, 77, 79, 80, 81, 82, 91, 93, 104, 105, 108, 109, 149–51, 215
refrigeration (defined), 7
refrigeration circuits, practical, 54–5
refrigeration cycle, theoretical, 13–14
refrigeration system service, 188–98
reheat coils, 113, 114
relative humidity, 14, 61, 66, 97–8, 111, 112–13
relay tester, 200, 202
resistance, 19–20, 72
resistance start, induction run (R.S.I.R.) motors, see under single-phase a.c. motors
reverse-cycle units, 64–5
rolling piston compressors, 50–1
roof fans, 134
room air-conditioners, 28, 59–65, 67–9, 207, 208
room temperature, 112
rotary compressors, 37, 50–1, 75
rotating vane compressors, 51
rotors, 25, 26–7, 50, 76;
  see also stator-rotor units
rupture discs, 81, 83–4

safety precautions/controls, 35–6, 45, 62, 63, 76–7, 123–4, 138, 188, 199;
  see also earthing; pressure relief devices

229

scale, 43, 44, 81, 83, 106, 113–14, 123
screw compressors, 38, 74–5, 76, 121, 122
sealed system pressure relationships, 11–12
secondary refrigerants, 43–5
sensible heat, 7, *8*, 14, 60, 103
shaded pole motors, *see under* single-phase a.c. motors
shaft seals, 42, 73–4, 191–2
sheet metal gauges, 216
sight glasses, 42, 73, 87, 88
single-phase a.c. motors, 23, 27–8, 61, 125, 204–6
   capacitor start, capacitor run (C.S.R.) motors, 28, *29*, 59, 61, 62, 202
   capacitor start, induction run (C.S.I.R.) motors, 28, *29*, 50, 56–7, 201, 202
   induction start, induction run (I.S.R.) motors, 27–8
   permanent split capacitor (P.S.C.) motors, 28, *29*, 61–2, 79, 202, 203, 204
   resistance start, induction run (R.S.I.R.) motors, *27*, 28, 50, 56, 201
   shaded pole motors, 28, *29*
slip ring starters, *see under* starters
solar energy, 116
soldering and brazing materials, 140
soldering/welding equipment, 140–1
solenoid valves, *see under* valves
specific gravity, 44
specific heat capacity, 6–7, 44, 213
split phase, 27
split system air-conditioners, 59, 65–6, 69, 111, 196
standard atmosphere, 9
star-delta starters, *see under* starters
starters, 25–7, *31*, 50
   across line, 207
   auto-transformer, 26, *76*, 207
   direct-on-line, 25–6
   part wind, 26, 73, 76, 121, 207
   reduced voltage, 75, *76*, 111, 121, 122, 207;
   see also individual reduced voltage starters
   slip ring, 26–7, *76*
   star-delta (YD), 26, 73, 76, 207
   variable resistance, 28–30
starting relays, 56–7, 201–2, 203
stator-rotor units, 30–1, 73, 74
stators, 25, 50;
   see also stator-rotor units
steel tubing, 215–16
suction-compression cycle, 48
suction line controls, 94–6
suction-line liquid separators, 150
suction lines, 41, 54, 65, 76, 105, 189
suction pressure, 13, 48, 50, 76, 90, 94, 95, 100, 107, 111, 189
sulphur candles, 41
synchronous speed, 25
system controls in refrigeration circuits, 55–7

temperature, 4–6, 7–9, 14, 20, 35, 56;
   see also dew point temperature; dry bulb temperature; effective temperatures; low temperature displays; low temperature refrigeration systems; pressure/temperature relationships; room temperature; wet bulb temperature
temperature scales, 4–6
test equipment, electrical, 199–201
test sequences, electrical, 201
thermocouple, *18*
thermostatic expansion valves, *see under* valves
thermostats, 30, 55–6, *57*, 59, 62, 63–5, 76, 80, 88, 110, 111, 112, 113, 125, 207
Thompson, William, 1
three-phase a.c. motors, 25–7, 79, 206–7
three-wire installations — d.c., 32
tic tracer, 199–200
torque, 23, 27, 28, 53, 61, 76, 90
transformers, 22–3
trichlorofluoromethane, *see* R11
two-stage (compound) compressors, 107–8
two-wire installations — a.c. or d.c., 32

vacuum, 1, 10–11, 142;
   see also pumps, vacuum,
valve plates, 47–8, 192–3
valves, 35–6, 59, 64–5, 76, 80–91 *passim*, 108, 144, 148, 195, 196–7
   constant pressure expansion (C.P.), 89–90
   discharge, 42, 47–8, 51, 73, 77, 108, 144, 196
   expansion, 13, 28, 76, 79, 89, 108, 124, 139;
   see also constant pressure expansion (C.P.) valves; thermostatic expansion valves
   hot gas bypass, 190, *191*, 196
   pressure relief, 50, 61, 81, 87, 119, 122
   solenoid, 21, 64–5, 76, 87, 88–9, 92, 93, 111, 124, 142, 176, 189–90
   suction, 42, 47–8, 73, 76, 108, 144, 196
   thermostatic expansion (T.E.V.), 90–1, 108, 110, 128–9, 188–9, 196–7
variable air volume (V.A.V.) systems, 132–3
variable resistance starters, *see under* starters
vee belt drives, 160–2

viscosity, 43
voltmeter, 19, 200
volts (V), 19–20
volumetric efficiency, 49

water chillers, 122–6;
  see also absorption water chillers; water chillers, packaged,
water chillers, packaged, 121–4, 207, 208
water-cooled condensers, 73, 80–2, 87, 111, 119, 122
  atmospheric, 81–2
  evaporative, 82, 86, 105, 137
  shell and coil, 81, 87, 111
  shell and tube, 80, *81*, 84, 86, 87, 105, 111, 118
  tube in tube, 81, 111
water cooling towers, 82–3, 87, 105, 137
water pumps, see under pumps
watts (W), 20
welded hermetic compressors, 2, 47, 71, 73, 76–7, 110, 121, 125, 144, 146, 194–5, 196;
  see also hermetic compressors
wet bulb depression, 14
wet bulb temperature (WB), 14, 82, 112, 197
window air-conditioners, see room air-conditioners
wiring diagrams, *56*, 62, *63*, *78*, *116*, 207
workshop layout, 197